KU-077-280

GENESCAPES

The ecology of genetic engineering

Stephen Nottingham

ZED BOOKS
LONDON & NEW YORK

Genescapes was first published in 2002 by
Zed Books Ltd, 7 Cynthia Street, London N1 9JF, UK and
Room 400, 175 Fifth Avenue, New York, NY 10010, USA

Distributed in the USA exclusively by Palgrave, a division of
St Martin's Press, LLC, 175 Fifth Avenue, New York, NY 10010, USA.

Editorial Copyright © Stephen Nottingham, 2002
Individual Chapters Copyright © Individual authors

Cover design by Andrew Corbett
Designed and set in 10/12 pt Monotype Baskerville
with Univers display by Long House, Cumbria, UK
Printed and bound in the United Kingdom
by Biddles Ltd, Guildford and King's Lynn

The rights of the author of this work have been asserted by him in accordance with the
Copyright, Designs and Patents Act, 1988

All rights reserved

A catalogue record for this book
is available from the British Library

ISBN   Hb 1 84277 036 5
       Pb 1 84277 037 3

Library of Congress Cataloging-in-Publication Data
has been applied for

Contents

Acknowledgements   vi
Abbreviations   vii
Glossary   viii

1. Introduction   1
   Genetic modification in perspective /1 What is genetic
   modification? /2 Differences in comparison to
   conventional breeding /4 Ecology and genetic engineering /6

2. The principles of ecology   8
   What is ecology? /8 From biosphere to individual /10
   Food webs and food chains /12 Mutualism /13
   Habitats and niches /13 Invasion ecology /16
   Species diversity /16 Defining ecological risk /17

3. Microorganisms   19
   Ice-minus /19 Bioremediation /21 Impacts on soil ecology /23
   Baculovirus bioinsecticide /28 Genetic exchange /29 Risk
   containment and assessment /30

4. Transgenic crops   33
   Herbicide-resistance /33 Insect-resistance /42
   Virus-resistance /45 The next generation /47

5. Invasion   54
   Species introductions /54 Predicting invasions /57
   Assessing the invasiveness of transgenic plants /58

6. Genetic pollution   62
   Gene flow from transgenic crops /62 Crop-weed hybridization /65
   Crop contamination: Impact on organic growers /75 Food
   contamination /78 Canola in Canada /80 Percy Schmeiser's
   canola /82 Seed contamination: The Advanta case /84

7. Impact on non-target species   87
   Natural enemies /87 Monarch butterfly /90 Farmland birds /94
   Bt toxins in the soil /96 Pollinating insects and honey
   contamination /97

8. Engineering solutions? 102

Promoters and tissue-specific transgene expression /102
Antibiotic-resistance marker genes /106   The manipulation of
flowering /109 Hybrid seed /111   Terminator technology /112
Apomixis /116 Chloroplast transformation /117

9. Trees 121

Increased growth rates /121 Low-lignin trees /122 Herbicide
and pest resistance /123 Orchard trees /124 Gene flow /125
Forest diversity /127 Transgenic trees and climate change /129

10. Fish 132

Modifications and benefits /132 Physiological costs of growth
enhancement /135 Risks to ecosystems /136 The need for
transgenic fish /139

11. Assessing ecological risk 141

Risk assessment /141 Transgenes within genomes /143
Arthropods, birds and mammals /145 Farm-scale
evaluations /147

12. Diversity 156

Crop diversity /156 Corporate diversity and the patenting of
life /159 An appropriate technology? /164 The sustainable
alternative /167 Prospects for transgenic crops /169

13. The precautionary principle 173

Defining the principle /173 Reasons to be cautious /174 The
Biosafety Protocol /180 Landscapes and genescapes /182

Bibliography 185
Index 205

Acknowledgements

I would like to thank Stephen Young and Christine Reeves for reading the first
draft of *Genescapes* and offering numerous suggestions. Many thanks also to
Robert Molteno and everyone involved at Zed Books.

Abbreviations

ACNFP	Advisory Committee on Novel Foods and Processes (UK)		Institute for the Semi-arid Tropics (India)
ACRE	Advisory Committee on Releases to the Environment (UK)	IITA	International Institute of Tropical Agriculture (Nigeria)
AFP	Anti-freeze protein	IPM	Integrated Pest Management
APHIS	Animals and Plant Health Inspection Service (of the USDA)	IRRI	International Rice Research Institute (Philippines)
		IVEM	Institute of Virology and Environmental Microbiology (UK)
BP	British Petroleum		
Bt	*Bacillus thuringiensis*	MAFF	Ministry of Agriculture, Fisheries and Food (UK) (replaced in 2001 by DEFRA)
CaMV	Cauliflower mosaic virus		
CFCs	Chlorofluorocarbons		
CGIAR	Consultative Group on International Agricultural Research	NFU	National Farmers' Union (UK)
		NERC	Natural Environment Research Council (UK)
CIMMYT	Centro Internacional del Mejoramiento de Maíz y Trigo (Mexico)	NGO	Non-Governmental Organization
		NIAB	National Institute of Agricultural Botany (UK)
CIP	Centro Internacional de la Papa (Peru)	NIH	National Institutes of Health (USA)
CO_2	Carbon dioxide	PBRs	Plant Breeders' Rights
CSIRO	Commonwealth Scientific and Industrial Research Organization (Australia)	PCBs	Polychlorinated biphenyls
		PROSAMO	Planned Release of Selected and Modified Organisms (UK)
DEFRA	Department for the Environment, Food and Rural Affairs (UK)	RAFI	Rural Advancement Foundation International (Canada)
DETR	Department of the Environment, Transport and the Regions (UK)	RSPB	Royal Society for the Protection of Birds (UK)
		SCIMAC	Supply Chain Initiative on - Modified Agricultural Crops (UK)
EC	European Commission		
EU	European Union	SCRI	Scottish Crop Research Institute
EPA	Environmental Protection Agency (USA)	SIT	Sterile insect technique
		TRIPs	Trade Related Intellectual Property Rights
FAO	Food and Agriculture Organization (UN)	UK	United Kingdom
FDA	Food and Drug Administration (USA)	UN	United Nations
		USA	United States of America
FMV	Feathery mottle virus	USDA	United States Department of Agriculture
FOE	Friends of the Earth		
GM	Genetically modified	VAD	Vitamin A deficiency
GMO	Genetically modified organism	WHO	World Health Organization
ICRISAT	International Crops Research	WTO	World Trade Organization

Glossary

Abiotic Non-living. Describes the physical or chemical aspects of an organism's environment.

Adaptation The process by which an organism undergoes modification through natural selection so that it is more suited to its environment.

Agrobacterium tumefaciens A bacterium which causes crown gall disease. It transfers a tumour-inducing portion of its DNA, in the form of a plasmid, into a plant's genome. This natural form of genetic engineering can be exploited to transfer any foreign gene (transgene) into a plant.

Agroecosystem The organisms and abiotic factors that interact in a field or other agricultural unit. An agricultural ecosystem.

Allelochemical A substance produced by an organism that is toxic or inhibitory to the growth of organisms in different species.

Amino acid The basic subunit of a peptide, polypeptide and protein. Proteins are composed of various proportions of about twenty amino acids.

Ammonia A water-soluble and colourless gas. Nitrogen-fixing bacteria convert ammonia into nitrate ions, which are further converted into organic compounds in plant roots.

Anther The part of the male reproductive organ of a plant from where pollen is released.

Antibiotic Substance from a microorganism that destroys or inhibits growth of another microorganism. Antibiotic-resistance marker genes have routinely been integrated into transgenic organisms, so that they can be selected from non-modified organisms during their commercial development.

Bacillus A group of relatively large, rod-shaped bacteria, that is ubiquitous in soil and air.

Backcross A cross between a hybrid and an individual from one of its parent populations.

Bacteria Diverse group of microorganisms all of which consist of a single cell lacking a distinct membrane-bound nucleus.

Baculovirus An insect-specific virus, used as a bioinsecticide.

Biogeochemical cycling The movement of chemical elements between organisms and non-living components in the biosphere, e.g. the nitrogen cycle, the carbon cycle.

Biomass The effective weight of living organisms in a given environment or ecosystem.

Biome A major ecological region, containing a distinctive plant assemblage, e.g. tropical rainforest, desert, tropical marine biomes.

Biotechnology The use of living organisms for human purposes. The development of techniques for the application of biological processes to the production of materials for use in foodstuffs, medicine, or industry. Genetic engineering is one set of such techniques.

Biotic Relating to life. The aspects of an organism's environment that consist of other living organisms.

Biosphere The whole of the region of the Earth's surface, the sea, and the air that is inhabited by living organisms. The whole Earth considered as one ecosystem.

Bromoxynil A herbicide produced by Rhône-Poulenc under the brand-name Buctril.

Bt *Bacillus thuringiensis*. A bacterial pathogen of insects, used as a bioinsecticide. Genes for *Bt* toxins are integrated into insect-resistant transgenic crops.

Carnivore An organism that consumes living animals or parts of living animals.

Catalyst A substance that increases the rate of a chemical reaction without itself undergoing any permanent chemical change. Enzymes are highly specific catalysts of biochemical reactions.

Cell The structural and functional unit of most living organisms.

Chloroplast Organelle in plant cells, containing its own DNA and the green pigment chlorophyll, where energy is produced through photosynthesis.

Chromosome Structure carrying genetic information in a linear sequence.

Cleistogamy The production of flowers, often inconspicuous, which do not open and in which self-pollination occurs.

Colonization The entry and spread of a species into an area from which it was previously absent.

Community A naturally occurring assemblage of populations of interacting organisms, living within a defined area or habitat. Species occurring together in space and time.

Competition An interaction between two (or more) organisms or species sharing an environmental resource that is in short supply.

Cross-pollination The transfer of pollen from male part of flower of one plant, to female part of flower of another plant having a different genetic make-up.

Cultivar A variety of a species produced and maintained by cultivation. A cultivated variety.

Decomposition The breakdown of complex, energy-rich organic molecules to simple inorganic constituents.

DNA Deoxyribonucleic acid. The molecule carrying the genetic information.

Dispersal The spreading of individuals away from each other, e.g. offspring from their parents.

Ecology The study of the interrelationships between organisms and their environment.

Ecosystem A visibly recognizable and self-contained unit of nature. The biotic community and its abiotic environment.

Emigration The movement of individuals out of a population or from one area to another.

Enzyme A protein that acts as a catalyst in biological reactions. Each enzyme is specific to a particular reaction.

Eutrophication Enrichment of a body of water with plant nutrients, e.g. nitrogen or phosphorus.

Extinction The condition arising from the death of the last surviving individual of a species, group, or gene, globally or locally.

F_1 generation Generation of hybrids from the cross of two parents differing genetically from each other. The first filial generation.

F_2 generation Generation of plants produced from interbreeding of hybrids or F_1 generation. The second filial generation.

Fecundity The number of eggs, or seeds, or offspring generally, produced by an individual.

Fertilization Union of sperm (pollen) and egg to produce offspring.

Fitness The contribution made to a population of descendants by an individual, relative to the contribution made by others. Reproductive fitness.

Food chain An abstract representation of the links between consumers and consumed, e.g. plant–herbivore–carnivore.

Fundamental niche The largest ecological niche that an organism or species can occupy, in the absence of competition from others, given the behavioural, physiological and genetic constraints operating for that organism or species.

Fungicide A chemical compound used to kill fungal pests.

Gaia The Gaia hypothesis states that the surface of the Earth is regulated by the activities of life.

Gene   A unit of heredity, usually carried on a chromosome, and consisting of DNA. Typically carries a code instructing a cell to make a specific protein.

Gene flow   The consequence of cross-fertilization. The exchange of genes between different, although usually related, populations, for example via pollen transfer.

Gene pool   The sum of all genes within a population.

Genetic engineering   A technique for combining DNA from different sources, outside a cell, using enzymes and vectors, and integrating this recombinant DNA into an organism so that it is expressed, and passed on to its offspring.

Genome   The complete genetic code of an organism.

Genotype   The genetic information contained within an organism.

Glufosinate   The active ingredient (glufosinate ammonium) in Aventis's Basta herbicide.

Glyphosate   The active ingredient in Monsanto's Roundup herbicide.

Habitat   The place where an organism lives.

Herbicide   A chemical compound used to kill plant pests (weeds). A weedkiller.

Herbivore   An organism that consumes living plant material.

Hybrid   An organism derived from two distinct parental lines.

Hyphae   Filaments of fungi, which may form a loose network in soil.

Immigration   Entry of an organism to a population from elsewhere.

Inbreeding   Mating between closely related parents.

Insecticide   A chemical compound used to kill insect pests.

Interspecific   Interaction between individuals of different species, e.g. interspecific competition.

Intraspecific   Interaction between individuals of the same species, e.g. intraspecific competition.

Introgression   The introduction of genes from one species to another via interspecific hybridization, or from one member of the species into another where the donor is geographically or morphologically distant from the recipient. The process by which new genes are introduced into wild populations.

Invasiveness   Capacity to spread beyond the site of introduction and establish in new areas.

Landrace   A locally adapted variety of a crop plant.

Lectin   A plant protein that binds to certain sugars on cell surface, causing cell clumping (agglutination), which can have an antifeedant effect.

Lignin   Complex carbohydrate molecule deposited within cell walls of woody plants, especially trees.

Marker gene   A gene used to identify genetically engineered organisms. They have typically conferred resistance to antibiotics.

Microorganism   Any microscopic organism, including bacteria, viruses and microscopic fungi.

Migration   The movement of individuals or whole populations from one region to another.

Mitochondria   Organelle that is the site of energy production in a cell. Mitochondria have their own DNA.

Mutualism   An interaction between the individuals of two (or more) species in which both derive benefits.

Mutation   A sudden random change in the genetic material of a cell.

Mycorrhizae   A common mutualistic association between the roots of a plant and a fungus.

Nematode   Roundworms. Microscopic free-living forms of nematode are extremely numerous worldwide, particularly in soils where they play an important role in the breakdown and recycling of organic matter.

Niche An ecological niche is an abstract concept defining the limits of all important environmental factors, both biotic and abiotic, within which a species can maintain a viable population.

Nitrogen-fixation A process by which atmospheric nitrogen is assimilated into organic compounds in living organisms.

Nucleus The organelle in plant and animal cells where the majority of the genetic material is concentrated. The control centre of a cell.

Organelle A subcomponent of a plant or animal cell, e.g. nucleus, mitochondria, and chloroplasts.

Outbreeding Mating between genetically dissimilar parents. Also called outcrossing.

Parasitism An association in which one organism (the parasite) lives on or in the body of another (the host).

Parasitoid Insect with free-living adult stage, and larval stage that necessarily develops on or inside a host of another species. The host is killed before the adult parasitoid emerges.

Pathogen A disease-causing microorganism.

PCBs Polychlorinated biphenyls. Synthetic chemicals used in electronics industry. Manufacture has now largely stopped. Highly toxic and very persistent in the environment. They have accumulated in food chains.

Pest An organism that interferes with the well-being of human beings. Microorganisms, fungi, insects, plants (weeds), rodents or any other organism could be a pest.

Pesticide Any chemical compound used to kill pests, e.g. herbicides, insecticides, and fungicides.

pH A scale of acidity (1 to 7) and alkalinity (7 to 14).

Phenotype The observable characteristics of an organism.

Pheromone A chemical substance emitted by an organism as a specific signal to another organism of the same species.

Phloem Plant tissue in veins of plants that is responsible for transport of organic solutes.

Photosynthesis The utilization of the energy from sunlight to synthesize organic compounds from carbon dioxide and water.

Plasmid Small circular DNA structure in bacteria.

Pleiotropy When a gene has a simultaneous effect on more than one characteristic.

Pollination The transfer of pollen from the male to the female part of a flower.

Population Group of individuals of a single species.

Predation The consumption of one organism by another, in which the consumer first attacks the consumed when it is alive.

Predator An organism that consumes other organisms.

Prey An individual liable to be or actually consumed by a predator.

Promoter A DNA sequence that switches on a gene.

Protein Molecule consisting of long chain of amino acids. Some are enzymes, others have structural role.

Protozoa A group of unicellular or acellular animals. Widely distributed in freshwater and marine ecosystems.

Realized niche The portion of a fundamental niche occupied by a species in the presence of competitors and predators.

Recombinant DNA DNA containing genes from different sources that have been combined by genetic engineering.

Resource In ecology, water, space (e.g. nesting space), light, food or nutrients, or anything else that an organism consumes. Species compete for resources.

Roundup The brand-name of a herbicide, whose active ingredient is glyphosate, manufactured by Monsanto.

Roundup Ready The brand-name for crop seed that is genetically engineered to be resistant to Monsanto's Roundup herbicide.

Seed bank i) The population of viable dormant seeds that accumulates in and on the soil. ii) A collection of seeds and germplasm from a cross section of plants, in which seeds are stored for long periods, e.g. using liquid nitrogen.

Self-pollination The transfer of pollen from male part of one plant to female part of the same flower or another plant with the same genetic make-up.

Species A taxonomic category typically (but not always) containing organisms that can actually or potentially interbreed.

Species diversity An index of community diversity that takes into account both species richness and relative abundance of species.

Species richness The number of species in a community.

Survivorship The probability of a representative newly born individual surviving to various ages.

Taxonomy The study of the rules, principles and practice of classifying living organisms.

Toxin A biological poison produced by an organism, usually in the form of a protein.

Transgene A gene introduced into an organism's genome using genetic engineering, usually from another species.

Transgenic plant A plant that has been genetically engineered.

Tri-trophic Three trophic levels (e.g. plant–herbivore–predator).

Trophic level Classification of organisms in a community according to their feeding relationships. The first trophic level comprises green plants, the second herbivores, and so on.

Variety A distinct subtype of a given species.

Vector In genetic engineering, a molecule, often a bacterial plasmid, used to transfer DNA from one organism to another.

Virus A microorganism capable of independent metabolism and reproduction within a living cell, but inert outside a host cell.

Volunteer plant A plant growing in a field in the year after it was sown, as a result of seed being shed and remaining in the soil.

Weed A plant pest that grows where it is unwanted, often on cultivated land where it competes with crop plants for resources.

This glossary was compiled with the aid of the following sources: *Oxford Concise Science Dictionary*, 3rd edition, Oxford: Oxford University Press, 1996; *Larousse Dictionary of Science and Technology*, New York: Larousse, 1995; Begon *et al.*, 1990; Krebs, 1985; Rissler and Mellon, 1996.

1 Introduction

An organism's genetics and its environment are closely interlinked. A change in environmental conditions can lead to genetic change within a population over time, with some genes increasing in frequency and others declining in frequency across several generations. Those organisms having the genetic make-up to thrive best under the new environmental conditions will survive to reproduce, while others perish. Organisms therefore evolve to become better adapted to their environment, through the process of natural selection. However, life on Earth is being increasingly affected by the activities of just one species. Man (*Homo sapiens*) is changing the genetic make-up of many of the Earth's living inhabitants by altering the planet's physical and environmental conditions, and through selective breeding programmes.

Genetic modification in perspective

Mankind has been modifying plant and animal species since the dawn of agriculture, over 10,000 years ago. Farmers have made genetic improvements by selecting crops and livestock, with what they consider to be desirable characteristics, at each generation. This process of artificial selection has always been vital to the course of mankind's progress. The cultivation of crop plants and the domestication of animals dramatically alters the landscape, while providing a stable food supply; the foundation for the growth of civilization. The Industrial Revolution provided great benefits, but also led to environmental change, for example through many types of pollution, and eventually climate change.

A major advance in food productivity occurred in the middle of the twentieth century, when progress in plant breeding and the development of new synthetic pesticides led to the Green Revolution. Large crop yield increases helped to feed the world's rapidly growing population. However, there followed adverse environmental impacts, and yield increases levelled off. Further scientific advances were needed to increase agricultural productivity.

Genetic engineering is part of the new genetics, based on molecular biology and biotechnology. It offers both the promise of increased food

1

production and a means of manufacturing a range of medical and industrial products in an environmentally friendly manner. Genetic engineering is also known as recombinant DNA technology. It has the capability of producing organisms with novel genetic make-up. Genetic improvement is no longer limited by the constraints of artificial selection, because genes can be inserted into organisms without the need for reproduction or crossing. The world's genetic resources can now be put to fuller use.

Genetic engineering has rapidly transformed agriculture. The first lines of genetically engineered or transgenic crops were released commercially in 1996. By the year 2000, the global area of transgenic crops was estimated to be over 44 million hectares, an area almost twice that of the United Kingdom. Most transgenic crops have been grown in developed or industrial countries, although the proportional area in developing countries is increasing. A little over 30 million hectares of transgenic crops were grown in the USA in 2000, while Argentina grew ten million hectares, Canada three million hectares, and China half-a-million hectares. Soyabean accounted for 58 per cent of transgenic crops in 2000, followed by maize (corn), cotton and oilseed rape (canola). Herbicide resistance has been the characteristic most commonly engineered into transgenic crops (James, 2000). Genetically modified microorganisms are grown in vats to produce food ingredients, and medicinal and industrial products, while modified microbes may before long be released into the environment in a range of applications. Genetic engineering may soon also become widely used in other areas, including forestry and aquaculture.

Genetic engineering has the potential to modify the Earth for better or worse. There could be great benefits for mankind, but also environmental costs. In agriculture, productivity may eventually increase in an environmentally friendly manner. However, the 'first generation' of transgenic crops has not fulfilled its initial promise. Benefits have been reaped by multinational companies and industrialized farmers, but not by consumers and small-scale or subsistence farmers. Costs include the erosion of traditional farmers' rights, the increasing privatization of genetic resources, a reduction in crop diversity, and a range of possible environmental hazards. The massive scale on which genetically modified organisms (GMOs) are being released into the environment means that there is the potential for severe ecological disruption.

What is genetic modification?

Two different definitions of genetic modification are in use. The first is that genetic modification encompasses all genetic improvement made through human intervention, by whatever means. This includes conventional plant- and animal-breeding methods. This view takes into account the fact that techniques involving the new genetics form a continuum with previous methods of genetic improvement. It downplays the unique features of genetic engineering, and is particularly favoured by people promoting advances in biotechnology.

The second view is that the term should be restricted to cases of genetic modification involving genetic engineering. This definition excludes all the other genetic improvements made through human intervention, and hence overplays the uniqueness of modifications made using genetic engineering. It is mainly associated with those critical of the new biotechnology. However, it is what people generally understand when the terms 'genetic modification', 'genetically modified', or 'genetically modified organism' are used. Genetic modification is also now the term used in much legislation, particularly in Europe, that oversees genetic engineering. Therefore, 'genetically modified organism' (GMO) will be used throughout this book, and it will be used exclusively for organisms modified through genetic engineering. The genes inserted into GMOs will be referred to as transgenes, while GM crops and GMOs will also be called transgenic crops and transgenic organisms, respectively.

Genetic engineering, or genetic manipulation, is legally defined (in the UK) as:

> The formation of new combinations of heritable material by the insertion of nucleic acid molecules, produced by whatever means outside the cell, into any virus, bacterial plasmid or other vector system so as to allow their incorporation into a host organism in which they do not naturally occur but in which they are capable of continued propagation (Walker, 1995).

The key stages in this procedure are integration, expression and transmission. A gene is said to be integrated when it has been successfully inserted into an organism's genome; it is said to be expressed when it produces protein within that organism; and a transgene is said to be transmitted when it is passed to the offspring of that organism. The key tools of genetic engineering are enzymes, which are responsible for the cutting, pasting and copying of DNA (deoxyribonucleic acid), and vectors, molecules used to transport transgenes between organisms. These manipulations occur outside cells (*in vitro*).

Transgenes are never integrated into organisms on their own. They are placed into gene constructs, alongside promoters and marker genes. Promoters ensure that a transgene is expressed in the novel genetic environment of another organism. A promoter from the Cauliflower Mosaic

Virus (CaMV), for example, is typically used for this purpose. In nature, this promoter ensures the virus's genes are expressed in the plants it infects. Marker genes are necessary because only a small number of organisms successfully take up transgenes during the transformation process. In plants, antibiotic-resistance marker genes are commonly used for this purpose, to enable modified material to be actively selected from unmodified material. It is gene constructs that are integrated into the genomes of organisms. The technical details of genetic engineering are outside the scope of this book. However, the techniques are now well established and described in many textbooks (for example Nicholl, 1994).

Differences in comparison to conventional breeding

Norman Maclean, in a review of animals with novel genes, concluded that genetic engineering is 'both essentially similar to more conventional methods of generating genetic variation and yet singularly different' (Maclean, 1994). In this book, the focus will be on the singular differences that make GMOs unique. These differences are best appreciated by comparing genetic engineering with conventional breeding methods. In conventional breeding two organisms must be crossed, in order for their genes to combine and re-assort. This process is constrained by the ability of two organisms to interbreed. With genetic engineering, novel genetic combinations are possible because a gene from one species can be inserted into widely differing species. Furthermore, two entire genomes are involved in conventional breeding, so that the genes of interest plus the rest of the genomes recombine. With genetic engineering, however, only the genes of interest are introduced. Genetic engineering is an advance, therefore, because the potential gene pool for selecting desirable genes is massively increased, and the desired outcome can be arrived at more quickly and efficiently.

It should be noted that several modern plant-breeding advances that generate genetic novelty are excluded by the above definition of genetic engineering, even though they extend the scope of genetic improvement in comparison to more traditional methods. For example, plants created by zapping cells with radiation to increase their rate of mutation (a truly mutant plant by any tabloid editor's definition) or by protoplast fusion (a method in which cells from differing species are stripped of their cell walls using enzymes, and fused to create novel genetic combinations) both fall outside the definition of genetic engineering. These methods do not involve the manipulation of DNA outside cells using enzymes and vectors. In addition, some of the characteristics associated with GMOs have also been introduced via conventional breeding methods. Insect-resistant crops, for example, have been produced using both conventional breeding and genetic engineering.

The most obvious feature, which singles out GMOs from other modified organisms, is that genes can be moved from one species to another. The species barrier has been broken. Critics of genetic engineering have focused on this fact, but some notes of caution need to be sounded. GMOs contain 'novel combinations of DNA', but this does not necessarily mean that they contain genes from another species. Gene-silencing manipulations, for example, switch off or silence existing (endogenous) genes. The manipulation of existing genes, via modifications to the genetic elements that control their activity, is likely to become an increasingly important area of genetic engineering. However, to date all GMO releases to the environment have incorporated genes from other species (even the gene-silenced GMOs have contained alien promoters and marker genes).

The phrase 'breaking the species barrier' has acquired connotations beyond its biological significance. In some quarters it is seen as going against nature in a very fundamental way. There are three points to make in response to this. First, all methods of genetic improvement, not just genetic engineering, involve putting genes where they probably would not have occurred through natural selection. Second, species are part of a classification system erected by humans. Classification is an effective but simplified model of the real world. It is subject to continual change, with taxonomists forever renaming species or reassigning them to different genera, families, and so forth. This often reflects the situation in nature, because evolution is still going on. Third, the species barrier is constantly being broken in nature. Genetic exchange occurs between non-interbreeding species, for example, via the mediation of viruses and bacteria. Some important plant viruses, for example, are natural genetic engineers that insert their genes into the genomes of plants. Genetic engineers exploit such natural systems (enzymes and vectors) in order to get genes from one species to function in another species. Genetic engineering is made possible by the universality of DNA. The genes that are responsible for the same functions in widely differing species are often very similar in terms of their genetic code. The extent of genetic exchange between species is only now becoming apparent. Nevertheless, the movement of genes between species is the basis for many of the concerns about the environmental impacts of GMOs.

This book is about the ecological risks, both actual and potential, associated with the release of GMOs into the environment. I will not be arguing that all applications of genetic engineering should be banned, that all genetic engineering is immoral, or that mankind should not be 'meddling with nature'. Many applications of genetic engineering will bring great benefits, while in other cases the various risks and costs will outweigh any benefits. Each case should be judged separately. Human beings are part of nature, and human ingenuity, be it rubbing two sticks together to start a fire or transferring a gene from a bacterium to a plant, comes from human nature. Genetic engineering is not immoral *per se* but it can become immoral if, for example, animals suffer as a consequence. The main aim of the book is to explain why releases of GMOs into the environment raise genuine ecological concerns.

Ecology and genetic engineering

The science of ecology is central to this book. Ecology is the study of organisms in relation to their environment, and is distinct from the wider areas of environmental science and environmentalism. It is important to understand the basics of the science of ecology, in order to make an informed view of the potential environmental impacts of GMOs. The possible environmental costs can then be balanced against the proposed benefits. A key theme will be how organisms are interconnected within ecological systems. If a gene is introduced which modifies the physiology or behaviour of an organism, then that modification can be expected to have impacts throughout an ecological system. Ecological principles can help to predict the potential impacts of a wide range of past, present, and future GMO releases to the environment.

The unique ecological risks associated with a range of GMOs are explored in this book. Microorganisms pose particular risks, because they are effectively irretrievable once released, multiply rapidly, and have a high capacity for genetic exchange. The book also considers the risks posed by transgenic trees and fish. The main focus, however, will be on transgenic crops, because of all GMOs they have been released into the environment on the largest scale.

Most transgenic crops have been modified for herbicide resistance, which enables weedkillers to be used more efficiently, followed by resistance to a range of pests and diseases. This 'first generation' of transgenic crops has provided benefits for growers on modernized farms, but they may exacerbate some of the ecological problems associated with high-input agriculture.

A number of ecological risks are associated with transgenic crops. They could turn out to be more invasive than equivalent non-transgenic crops, for instance becoming weeds in their own right. More probably, they could cross with wild relatives in and around agricultural land to produce hybrid offspring containing transgenes, which could become established in wild populations. Pollen from transgenic crops can also fertilize non-transgenic cultivars, resulting in problems for organic growers and producers of seed crops. Unwanted contamination of food can result. We may be approaching a situation where a degree of 'genetic pollution' becomes commonplace due to the cultivation of transgenic crops.

Transgenic crops were designed to have specific effects on a limited number of organisms, for example a particular pest species. However, a range of direct or indirect effects on non-target organisms, including beneficial insects, soil microorganisms, and farmland birds, can occur. Ecological interactions are complex, and adverse environmental impacts can be felt along food chains and throughout ecosystems.

There are a number of technical solutions for reducing the ecological risks associated with transgenic crops. Risks associated with promoters and antibiotic-resistance marker genes, for example, can be addressed by replacing them with alternatives. Sterility can be used to prevent transgene escape in pollen or seeds. However, 'technological fixes' do not address fundamental questions about the desirability of transgenic crops, and can raise their own socio-economic concerns.

Sustainable agricultural methods that embrace diversity provide an alternative to the centralized and industrialized agriculture that has surrounded transgenic crops to date. In many cases for which they are proposed, transgenic crops might not be the most appropriate technology to deploy. If transgenic crops are to bring benefits to the developing world, they may need to be developed in a much more sustainable and 'people-friendly' manner.

The accumulating evidence of ecological risk from studies on a wide range of GMOs, including microorganisms, insects, plants, and vertebrates, should serve as a warning to those regulating releases to the environment. There is a strong case for proceeding with more caution. Legislation that uses a framework based on the precautionary principle could ensure that risks are reduced. But strong commercial pressures continue to propel biotechnology forward at a rapid rate. Meanwhile, unpredictable ecological effects cannot be legislated for. The complexities of genetic architecture and ecosystem structure should not be underestimated.

2　The principles of ecology

Ecology is the study of organisms in relation to their environment. The word 'ecology' was first defined by a German zoologist called Ernst Haeckel in 1866, who derived it from the Greek for 'home' (*oikos*) (Haeckel, 1870). The Earth is home to all the organisms that live upon it. Ecology is the study of the relationships of those organisms to their environment and to each other. This chapter provides a brief introduction to the concepts and principles of ecology.

What is ecology?

Ecology is concerned with the interactions between living (biotic) and non-living (abiotic) things; with how organisms utilize the resources of the Earth and how these resources cycle within living systems; and with all the possible interactions that determine the distribution and abundance of organisms. Ecology is a scientific attempt to understand nature as a whole, and to unravel all its interconnections. In this book, an ecologist is a research scientist who studies organisms in relation to both the biotic and abiotic components of their environment, and not a green activist or a politician.

Ecology grew out of an experimental approach to studying natural history. Ecological principles emerged as studies accumulated and patterns were observed. The basic concepts of community and ecosystem were established in the late nineteenth century; the concept of plant succession emerged in 1916 (Clements, 1916); the understanding of how food chains influenced population size and community structure dates from the 1920s (Elton, 1927); and the concept of the biosphere from 1929 (Vernadsky, 1929). The early ecologists were influenced by the work of Charles Darwin, and evolutionary considerations are central to ecological thought to this day. The work of R.A. Fisher in the 1930s set the stage for the modern fusion of evolution and ecology. He provided a mathematical basis for considering the processes of natural selection and population change (Fisher, 1930). Bill Hamilton advanced this mathematical approach in the 1960s. His genetic explanation of kin selection, whereby the behaviour of individuals could be predicted by how many genes they shared in common, led to the development of sociobiology (Hamilton, 1963; Wilson, 1975).

Ecology is a quantitative discipline, in which experiments on constituent parts of a system and holistic approaches go hand in hand. Ecologists conduct experimental studies designed to test a particular hypothesis, for example whether transgenic pollen harms honeybees. They are also concerned with how entire ecological systems function. This holistic approach is called systems analysis and involves the study of biogeochemical cycles. Elements which are vital to life processes, such as carbon (C), nitrogen (N) and phosphorus (P), are traced as they cycle from the abiotic (non-living) environment through living systems and back to the abiotic environment. If a genetically modified organism (GMO) were to alter the availability of an element in the environment, for example nitrogen in the soil, then its impact might be felt throughout an ecological system via that element's biogeochemical cycle.

The systems approach complements the tightly focused experimental approach. They are top-down and bottom-up ways of looking at ecological systems. The dichotomy can be misleading, however, as the two approaches are closely integrated. The systems approach needs to be applied with rigour, needs the input of realistic data from natural systems, and needs to address real scientific questions. It stops well short of the holistic analysis of many political or philosophical environmentalists. Ecology is distinct, for example, from 'deep ecology', which emerged in the 1970s and is associated with the philosopher Arne Naess. Fritjof Capra has written that: 'Ultimately, deep ecological awareness is spiritual or religious awareness' (Capra, 1996). The science of ecology is not concerned with the mystical properties of systems. Incidentally, some 'deep ecologists' have argued that scientific ecology views humans as above or outside nature. This is untrue; an understanding of ecological systems requires that the role of humans *within* those systems is understood. The influential Gaia hypothesis, which considers major features of the Earth's surface (e.g. temperature, chemical composition) to be regulated by the combined action of life on the planet, can be both studied by ecologists and embraced by 'deep ecologists' (Lovelock, 1988). Systems analysis can ask specific questions about feedback mechanisms, for example, involving populations of algae in the oceans and how they influence levels of atmospheric chemicals. Philosophers can use Gaia fruitfully in formulating a holistic view of the world.

Ecology has a false image of being a soft or backward science. This view was typified in a highly misleading BBC documentary, broadcast in Britain in the early 1990s as part of the *Horizon* series ('Guess What's

Coming to Dinner?'), in which ecology was described as, 'a relatively undeveloped science using unsophisticated, almost nineteenth-century tools'. Ecologists were depicted in the field sampling seedlings in a study of the possible invasion into natural habitats of transgenic oilseed rape plants. Basic field equipment still works as well as it always has so it is no surprise that some of it resembles that used by nineteenth-century naturalists. Sampling for insects still involves using sweep nets, while plants are counted within simple squares called quadrats (Southwood, 1978). However, today these field-sampling methods are just the most visible end of a science that involves state-of-the-art computing, sophisticated mathematical modelling, and cutting-edge genetic analysis techniques. Ecology is a fast-moving and exciting science, as modern as any other. It is intellectually fascinating and has important applications. As mankind modifies the Earth to an ever greater extent, the importance of ecology will grow.

A number of excellent textbooks describe the history, development and principles of ecology (e.g. Begon *et al.*, 1990; Brewer, 1994; Collinvaux, 1993; Krebs, 1985; Owen, 1974; Ricklefs, 1973). These can be referred to for treatment of the subject in greater depth. The remainder of this chapter will be devoted to a consideration of some of the concepts and principles of ecology that are particularly relevant to the release of GMOs into the environment.

From biosphere to individual

Ecologists study organisms on a number of different levels: biosphere, biome, ecosystem and community, population, and individual. Each of these levels is a sub-set of the previous one.

Biosphere   The biosphere is the whole Earth considered as one ecological system. Biospheric studies are concerned with the cycling of minerals, nutrients and gases through living systems on a planetary scale. Few studies have considered the impact of GMOs at the biosphere level, although the release of a genetically engineered microorganism has raised concerns about possible impacts on the climate (see Chapter 3).

Biome   The biosphere is composed of several types of biome, consisting of major vegetation types, such as the arctic tundra or the tropical rainforest. A biome is a plant–animal formation of geographical extent. These are usually too large in themselves to be the subject of ecological study. The human impact on climate at the biosphere level may soon affect terrestrial biomes, particularly through altering the ranges of key vegetation species.

Ecosystems and communities An ecosystem is a subsection of a biome. It has biotic and abiotic components. Each ecosystem has a distinct set of interacting organisms. Ecosystems are self-sufficient units and represent the smallest unit within the biosphere containing all the characteristics to sustain life. Biologically important elements (e.g. C, N and P) cycle much more rapidly within ecosystems than between them. Changes in one part of an ecosystem can affect other parts, for example via the food chain. Ecosystems, such as ponds and forests, have been the subject of many ecological studies. An agricultural ecosystem or agroecosystem is a distinct cultivated ecosystem.

A community is a group of species that occur at the same place and at the same time. Communities are composed of populations of species engaged in various interactions. Communities of organisms, for example, may occur on a particular crop in an agroecosystem. A community plus its habitat can often be considered as an ecosystem. In ecosystem and community-level systems, everything is interconnected.

Population A population is made up of individuals of the same species that occur in the same place and at the same time. Populations naturally fluctuate, through variations in birth and death rates. Populations also rise and fall through the processes of emigration and immigration. Emigration occurs either through dispersal, the movement of individuals away from their home site (e.g. plant seed in the wind and plankton in ocean currents), or migration, the directed movement of individuals or sometimes whole populations from one region to another (e.g. monarch butterflies and many bird species). Most dispersal does not change the geographic range of a species, but occasionally dispersal contributes to a species' range expansion. The density of a population may act as a regulating factor in population size; for example, in crowded conditions competition for resources might act to decrease population size, by decreasing the birth rate or increasing the emigration rate. When assessing the ecological risks of GMOs it is important to determine how populations increase and disperse, in comparison to populations of comparable non-GMOs.

Individual An individual is a single organism. It can be studied with respect to its own species (intraspecific interactions) or different species (interspecific interactions). Competition for resources, for example, can be intraspecific or interspecific. A range of abiotic and biotic factors affect whether an individual survives to reproduce and how fecund it is. Fecundity is a measure of the number of offspring each individual produces. If an individual's genes are passed to many offspring before it dies, it is said to have a high reproductive fitness. Evolution acts on individuals and their genes, with the best-adapted individuals contributing more genes to future generations. The relative reproductive fitness of individuals is an important consideration in assessing the ecological risk posed by a GMO.

Food webs and food chains

Most of the principles relevant to the environmental release of GMOs are concerned with the ecosystem and community level. A food web, for instance, is a diagram depicting the interaction of species within a community. It shows what resources each species in a community needs to grow and reproduce, or who eats what. Food webs are usually highly simplified representations of natural systems, but they are central to an understanding of the processes at work within ecosystems.

A single sequence of feeding interactions in a food web is called a food chain. Food chains can be considered in terms of trophic (literally, 'feeding') levels. A trophic level is the position of an organism in the food chain, assessed by the number of resource-transfer steps to reach that level. The main trophic level categories are producers, consumers, and decomposers. Producers lie at the bottom of food chains and make organic compounds (food) from simple inorganic materials; consumers obtain their food by eating other organisms; while decomposers use dead plant or animal material as a food source. In terms of numbers, there are usually more producers than primary consumers (herbivores or plant-eaters), which in turn are more numerous than secondary consumers (carnivores or animal-eaters). This gives a pyramid of numbers that is characteristic of a particular community's structure. It reflects how food resources flow through a community. A typical food chain, therefore, may consist of a plant (primary producer), a herbivore that consumes that plant (primary consumer), a predator that consumes that herbivore (secondary consumer), and microorganisms that decompose organisms from the other trophic levels. A change at one trophic level can have repercussions at another level (Pimm, 1982).

A genetic modification to a plant could therefore affect a predator via a physiological or behavioural change in a herbivore. This effect, which is felt along a food chain, is called a cascading trophic interaction. The potential for transgenic crops to cause indirect environmental impacts via such interactions has been demonstrated in experimental studies (Chapter 7).

Mutualism

Another relationship between species, distinct from plant–herbivore or predator–prey interactions, is one of mutualism. This is an interaction between two species from which both derive benefit. This is also referred to as a symbiotic relationship.

Mycorrhizae   A particularly important mutualistic association occurs between the roots of plants and certain fungi, which form mycorrhizae (literally 'fungus-root'). Mycorrhizal fungi are vital to plants because they aid in mineral and nutrient uptake. Mycorrhizae also enhance a plant's ability to withstand water stress and disease. In return, the fungi benefit by deriving photosynthetic products from plants, on which they depend for food and energy. The extensive network of filaments (hyphae) produced by the fungi can extend for several metres, greatly increasing the volume of soil from which a plant can obtain resources. Mycorrhizal fungi can form a vast underground network, which can turn a plant community into a single functional unit. In the case of trees, the extensive mycorrhizal network can in effect make a forest an environment of considerable inter-connection, as the fungal hyphae from different tree species intertwine in the soil and exchange mineral resources (Woods and Brock, 1964). Many plants simply cannot survive without mycorrhizal fungi. It has been said that plants do not strictly have roots; they have mycorrhizae (Brewer, 1994). It is difficult to overstate the importance of mycorrhizae. If a GMO were to have a negative effect on mycorrhizal fungi, the impacts on soil ecology could be considerable. Laboratory studies have demonstrated such effects (Chapter 3).

Pollinators   Pollination is an important symbiotic interaction between plants and insects, birds or mammals. Flowering plants produce food resources for insect pollinators, for example, while insects transport pollen from male to female flower structures, thereby fertilizing them. Pollinating insects are a vulnerable group of non-target species in fields of transgenic crops. If transgenic crops were to harm honeybees, for example via an insecticidal protein expressed in their pollen, the pollination of a wide range of crops and native flowering species could be affected.

Habitats and niches

Habitat A habitat is a physical place that provides the environmental con-ditions and resources required by a species. It is a favourable environment, and often one that an organism actively selects. It is broadly defined as the specific set of environmental conditions under which an individual, species, or even a community exists. It is a species' address, or the location where an organism can be found. A species' habitat is distinct from what ecologists call its niche.

Niche   A niche is a species' particular role within its community. It is not a physical location, but an abstract concept describing all the environmental conditions required for a species to maintain a viable population. The niche of a species is determined by all the parameters that restrict a species' range, for example the temperatures within which that species can grow and reproduce. Many other abiotic factors, along with interspecific competition and predation, determine a species' niche.

Fundamental niche   Evelyn Hutchinson formulated a precise mathematical definition of a niche (the n-dimensional hypervolume), which envisaged a species occupying an abstract space (niche space) comprising axes (n in number) for each important parameter. If a species can establish a population only within a particular temperature range, for example, then that temperature range forms one axis. The three most important axes may be its temperature range, the size of food items it can physically eat, and the humidity levels it requires to survive. The volume within these intersecting ranges would give a simple three-dimensional model of a niche. A many-dimensional figure can be theoretically constructed in this manner within the upper and lower limits of each significant parameter. Ecologists refer to this as the fundamental niche. It can have an infinite number of dimensions (remember this is a mathematical model of reality), and represents what a species is capable of, given the physiological, historical (genetic) and behavioural constraints on individuals of that species (Hutchinson, 1957).

Realized niche   A niche is therefore a property of a single species, and a fundamental niche indicates a species' potential role in an ideal world. In a community, however, many species co-exist. Therefore, fundamental niches are rarely if ever realized. Niches are maintained through competition between species for limited resources. A competitive species can successfully exclude another species from utilizing a particular resource that it would otherwise have exploited. Competing species have niches that partially overlap, and the niche space boundary of a less successful competitor can be pushed back. The niche space bounded by the actual ranges of each parameter found in nature is called a realized niche. A species' realized niche is smaller than its fundamental niche owing to the presence of interactions between species. Ecologists can calculate niche separation distances and niche overlaps between species from field-collected data to build up a picture of community structure.

There is a limit to how much niche overlap can occur because two species cannot occupy the same ecological niche. In other words, complete competition cannot exist between species. This is known as the competitive exclusion principle (Hardin, 1960). If interspecific competition becomes too severe, then either selection pressure will work to develop species differences that allow for continued co-existence (e.g. a genetic change occurs in one species that enables it to exploit a new resource) or one species will become extinct. If a species disappears from a community, ecologists can predict how the other species in the community may react from an understanding of their niches.

Niche and GMOs Niche theory can also be used to predict how GMOs might behave on release into the environment. The introduction of a transgene can affect a species' ability to exploit resources, thus changing its niche. When niches are altered, ecological impacts can be expected. In extreme cases, this could involve changes in community structure and bio-diversity. Genetic modification can alter the fundamental or realized niche, or both. If the realized niche is altered, then a GMO can become a better competitor within its native or home range. It does what its parent (equivalent but unmodified organisms) can do, but better. This could occur, for example, as a result of a GMO having an increased growth rate or larger size. A significant change in its realized niche may confer substantial advantages over unmodified individuals. It may even drive parent populations towards extinction. Changes in realized niche could also alter inter-specific competition, making a GMO a better competitor and enabling it to invade other species' niche spaces.

When the fundamental niche is altered, a species can escape from the native or parental range due to an ability to obtain novel resources. In the case of a GMO, a change in fundamental niche can be said to have occurred if it expands its range or exploits a resource beyond the means of an equivalent unmodified organism. Added characteristics, like freeze-tolerance in fish, could enable GMOs to exploit novel resources, owing to an extended range. In terms of assessing ecological risk, slight changes in realized niche present very low risk, significant changes in realized niche present relatively high risk, while changes in fundamental niche present potentially very high risk (Gutrich and Whiteman, 1998). If GMOs enter a new range, many of the constraints on individuals, for example those imposed by competitors or predators, are no longer present, and they could become invasive.

Invasion ecology

The dispersal and expansion of a species into a new range is called an invasion. It represents a sudden exploitation of new niche space. A species can become invasive when it finds itself in a new and favourable environment, but without the predators and parasites that control its numbers in its home range. Invasions of exotic species can occur naturally or as a result of human intervention, which may be inadvertent or deliberate. Many cases of accidental introductions of exotic species have been recorded, and in some cases these species have become major pests. Biological control is the release of a predator, parasite or pathogen into a new range to control a prey species that may have already expanded its range to become a pest. Biological control is therefore in itself risky, as an introduced species could itself become an invasive pest.

Genetic modification can alter the characteristics of an organism so that its chances of becoming invasive are increased. The behaviour of known invasive species can be used as a model to assess the potential invasiveness of GMOs (Chapter 5).

Species diversity

Measuring diversity   Species diversity is an important descriptor of an ecological community. The commonest measure of diversity is to count the number of species per unit area. This gives an indication of the total number of species or the species richness. However, this does not take into account the relative abundance of each species; therefore a number of other indices have been devised to do that. For example, the Sequential Comparison Index takes into account the number of individuals of each species in relation to the total number of individuals in the sample. The various diversity indices can be compared for a range of communities sampled (Pielou, 1974). Biodiversity is generally considered to be a measure of the number of species and their relative abundance.

Biodiversity   Globally, biodiversity changes as a result of the processes of species formation (speciation) and extinction. Biodiversity decreases when the rate of speciation is lower than the rate of extinction. On a local scale, biodiversity is determined by colonization and local extinctions. Island biogeography theory has led to an understanding of how species number changes with increasing island area, increasing isolation distance, and over time (McArthur and Wilson, 1963). The findings are applicable to any distinct ecological unit, not just islands. They have relevance, for example, to the conservation of endangered species and the design of nature reserves. Biodiversity is generally highest in the tropics, where climate is warmer and subject to less seasonal fluctuation. Increased diversity in a habitat is generally accompanied by an increase in specialization, with the niches of species becoming narrower and more tightly packed.

Human intervention Human activity has dramatically reduced biodiversity globally, mainly through habitat destruction. Ecologists have calculated that extinction rates are on the increase. The reasons for preserving biodiversity fall into three main areas. First, it is important to preserve genetic resources. Second, a decline in species could irreversibly disrupt the biogeochemical cycles that support all life. One theory suggests that the disappearance of certain key species (keystone species), which have a disproportionate influence on these cycles, could send cascading effects through an ecological system. Third, it is ethically unacceptable to wipe out species if we can avoid doing so, while wildlife should also be preserved on aesthetic grounds. For these reasons, it is important to safeguard existing levels of biodiversity (Wilson, 1992). The introduction of GMOs could potentially impact on biodiversity in a number of ways, which are explored in later chapters.

Defining ecological risk

Ecological risk should be assessed before GMOs are released into the environment. Ecologists have therefore to define what they mean by an ecological risk in the context of GMOs. An ecological threat can be said to exist if the presence of an organism containing a transgene results in a negative impact on an ecological system, usually an ecosystem (Hails, 2000). To say whether there is risk, ecologists need to make comparisons, with and without a GMO, or before and after the introduction of a GMO. This comparison with the existing situation is particularly important in agricultural ecosystems, as modern farming methods have already had a large impact on biodiversity. However, manipulation experiments of this type are few and far between. Most of the experiments done with GMOs are laboratory-based or small-scale field studies where no ecological data has been collected (Chapter 11). Nevertheless, a larger picture is starting to emerge, from which a framework for assessing risk can be erected.

The accumulation of data from experimental studies and the application of ecological principles, such as invasion and niche theory, enable certain predictions to be made about the ecological risks associated with

GMOs. The evidence to date shows that even though the risks in most cases are relatively small, there is potential for a wide range of direct and indirect ecological effects. Once spread to the wider environment, GMOs, or the transgenes originating from them, may be impossible to eradicate and may affect a number of other organisms. Identifying ecological risks at an early stage is therefore important. If ecological risk is missed or underestimated, there could be significant consequences throughout ecological systems.

3 Microorganisms

Microorganisms have been used in biotechnology for centuries to produce foodstuffs, such as bread, cheese and beer, and medicinal and industrial products such as antibiotics. Genetic engineering is extending the possibilities of these applications, which usually take place in fermentation vats or other enclosed vessels. But the advent of genetic engineering has also led to many proposed applications involving the deliberate release of modified microorganisms into the environment.

The potential ecological impacts, which are outlined in this chapter, could be considerable. Microorganisms can reproduce rapidly, disperse over long distances, and exchange genetic information freely. It would be very difficult to recover a strain of microbial genetically modified organism (GMO) that became established in the environment. If they dispersed, ecological effects could be felt over a regional, or even a global level.

Ice-minus

The first government-approved release of a GMO into the open environment anywhere in the world occurred on a small plot (less than 100 m²) at a University of California research station in April 1987. Steven Lindow and his colleagues here released a modified strain of a soil-dwelling bacterium called *Pseudomonas syringae*. Normal strains of this bacterium have proteins on their surface, under the control of a gene called *ice*, which initiate the formation of ice crystals. In the early 1980s, University of California researchers produced a genetically modified strain of *P. syringae* lacking these ice-forming proteins, called *ice-minus*. In laboratory experiments, *ice-minus* conferred frost-tolerance on strawberries on to which it was sprayed. Whereas frost damage occurred at 0°C with control strawberry plants, frost did not form on the leaves of plants sprayed with *ice-minus* until the temperature reached -5°C. The *ice-minus* sprays achieved this by competing with, and displacing, naturally occurring unmodified bacterial strains on the plant's leaves, flowers and roots. In the absence of normal ice-nucleating proteins, ice crystals form at a lower temperature and frost damage to fruit is reduced.

In 1984, Advanced Genetic Sciences (AGS) applied for permission to release *ice-minus* in California. Approval was given in 1986, after a court case in which the judge said that environmentalists had failed to show that harm would result from the field trial. Opponents of the trial feared that *ice-minus* might persist in the environment and have adverse ecological or even climatic effects. The ecologist Eugene Odum, for example, articulated concerns that the modified bacterium might affect ice-formation processes in the biosphere (Odum, 1985).

Monitoring during the initial field trial showed that *ice-minus*, which was applied as an aerosol, remained close to the target plants. Although the speed and direction of the wind had an effect on its dispersal, it was estimated that only 0.001 per cent of the bacteria released reached the edge of a 15-metre uncultivated buffer zone around the experimental plot. In a subsequent trial, with potato plants in varying weather conditions, some viable *ice-minus* were detected 30 metres from the test plants. *Ice-minus* persisted in a viable state in the soil for about a week after spraying (Lindow and Panopoulos, 1988). However, modified bacteria could have spread beyond the test site and not been detected by the monitoring methods used. The modified bacterium helped protect plants against frost damage, although not to any spectacular extent, and no adverse ecological effects were reported. Nevertheless, if *ice-minus* were sprayed over large areas, the potential risks would be much greater, especially if wind-borne bacteria were carried up into the atmosphere. Since the initial *ice-minus* trial, a greater appreciation has arisen of the possible ecological risks that such releases could pose.

It is now thought that microorganisms exert a significant influence on weather patterns. Bill Hamilton, of the University of Oxford, and Tim Lenton, a chemist at the University of East Anglia, both in England, developed this idea in the late 1980s, as part of a scientific investigation into the Gaia hypothesis. The Gaia hypothesis, as formulated by James Lovelock, states that living organisms regulate conditions on the Earth. In other words, life has not merely evolved to the conditions found on Earth, but the temperature and chemical composition of the planet are dynamically regulated by life. This is achieved through feedback mechanisms, analogous to the way cooling mechanisms are triggering in an overheated organism. Without life, the Earth would therefore revert to a very different set of conditions, comparable to other planets in the solar system, which would support little in the way of life. The production of chemicals by marine microorganisms, for example, influences the chemical composition of the atmosphere. Cloud formation can be seen as part of this regulatory system. Clouds modify the amount of solar radiation reaching the planet's surface, and where and how much precipitation falls in particular regions.

Clouds cover around 60 per cent of the Earth's surface, and microorganisms can be transported long distances within them. Hamilton and Lenton suggested that microorganisms in clouds could induce water condensation and ice formation, thereby influencing world climate. *P. syringae* is thought to be the most abundant airborne microorganism having natural ice-forming properties. It is therefore likely to play an important role in triggering rainfall or hail within clouds (Hunt, 1998; Hamilton and Lenton, 1998).

It is now known, thanks to research conducted high in the Austrian Alps by Birgit Sattler and her colleagues from the University of Innsbruck, that airborne bacteria are not merely transported, but can accumulate, grow and reproduce within clouds. They can form thriving colonies in cloud droplets, producing proteins and other products of microbial growth processes, at temperatures at or below 0°C. It was concluded that cloud water should be considered as a microbial habitat (Sattler *et al.*, 2001). If *P. syringae* were actively growing in clouds, its role in influencing climate may be even more significant than previously thought. Competition between natural strains of *P. syringae* and airborne *ice-minus* in the atmosphere could therefore affect cloud formation and rainfall patterns (Marchant, 2000).

Bioremediation

An area where microbial GMOs could play a particularly important role is bioremediation: the deliberate application of biological activity to reduce the toxicity of contaminants (Atlas, 1995). Plants and trees have also been used for bioremediation, but bacteria probably have the greatest potential for cleaning up contaminated land. Microorganisms can either biodegrade chemicals, by releasing enzymes that break them down into other compounds, or bioaccumulate them inside their cells. Unfortunately, in this increasingly polluted world, these are often slow processes.

Oil slicks   At its simplest, bioremediation involves the spreading of fertilizer to encourage the growth of naturally occurring bacteria that break down chemical contaminants. This has been done, for example, in response to major oil spills. When the *Exxon Valdez* grounded off Port William Sound in Alaska in March 1989, shedding its cargo of more than ten million gallons of petroleum, a fertilizer mixture, rich in nitrogen and phosphorus, was sprayed on the affected shoreline to encourage the bacterial breakdown of the oil slick. However, this could not prevent extensive mortality to birds and other wildlife (Prince, 1997).

A more sophisticated form of bioremediation is to seed contaminated areas with biodegrading or bioaccumulating microorganisms. Commercial bioremediation products, containing mixtures of microorganisms, are currently available for treating particular types of pollution incidents. However, the slowness of the process, the lack of single bacterial strains that can cope with more than one pollutant chemical, and the fact that bacteria can metabolize contaminants over only a relatively narrow range of concentrations, constrain the efficiency of existing solutions. Genetic engineering is now being applied to increase the efficiency of bioremediation.

The first patent granted to a life form was awarded to a novel strain of *Pseudomonas* bacteria, genetically engineered to digest oil slicks. In a landmark court case in the USA in 1980, the Supreme Court ruled that the bacterium was a human invention rather than a naturally occurring species (Chakrabarty, 1981). This opened the floodgates for the patenting of GMOs.

The use of genetic engineering to improve bioremediation could have many environmental benefits. In the case of oil spills, however, too much focus should not be placed on the release of GMOs to provide solutions. It would be better if oil spills did not occur in the first place; measures such as using double-hulled tankers or levying heavier fines on oil companies could go some way to ensuring this.

PCBs   In 1996, the Environmental Protection Agency (EPA) granted permission for a genetically engineered strain of *Pseudomonas putida* to be used to help clear up polychlorinated biphenyls (PCBs) in an enclosed industrial site in the USA. A gene introduced from another bacteria, *Alcaligenes eutrophus*, enabled the soil-dwelling *Pseudomonas* strain to break down the PCBs, which are major environmental contaminants. The PCBs are a family of synthetic chemicals that existed nowhere in nature prior to their introduction by the Swann Chemical Company in 1929 (Monsanto since 1935). PCBs are highly stable and non-inflammable, and became an important material within the electronics industry. However, the very stability of PCBs means that they do not readily break down, while biodegrading microorganisms have had little evolutionary time to adapt to them. By the 1960s, fish, birds and animals, in every ecosystem around the planet that ecologists examined, contained traces of PCBs. An estimated 3.4 billion pounds of PCBs had been produced by the time bans on their manufacture were first put in place in the late 1970s; although they are still found in electronic equipment and waste. They are known to be hormone disruptors, and may pose reproductive problems to wildlife, and possibly even humans, for centuries to come (Colborn *et al.*, 1996).

PCBs are a type of man-made pollutant ideally suited to treatment by GMO releases.

Strains of *Pseudomonas* have now been genetically engineered to degrade a range of chemical contaminants. For example, *P. putida* has been engineered to degrade 3–chlorobenzoate and 4–methyl benzoate in activated sludge and contaminated river-water sediments. Some strains of *Pseudomonas* have been engineered with multiple detoxification genes, for example to biodegrade compounds such as chlorofluorocarbons (CFCs) and PCBs simultaneously (Doyle *et al.*, 1995).

Herbicides Within agriculture, the fundamental niches of crop plant species are constrained by abiotic factors, including soil chemistry parameters such as nutrient and mineral levels, pH, and the presence of toxic chemicals and metals. GMOs are being designed to clean up contaminated soil, which could increase crop productivity on marginal land and reclaim once-productive agricultural land. The bacterium *Pseudomonas putida*, for example, has been genetically engineered to break down the herbicide 2,4–D (2,4–dichlorophenoxyacetate) in contaminated agricultural soils (Short *et al.*, 1990).

The modified *Pseudomonas* bacterium degraded 2,4–D into a compound called 2,4–DCP (2,4-dichlorophenol), however, which built up in the soil. Laboratory experiments conducted by the EPA, in the USA, demonstrated that this breakdown product was highly toxic to beneficial soil fungi, dramatically and irreversibly curtailing their growth (Doyle *et al.*, 1991; Doyle *et al.*, 1995). Although it is unclear whether 2,4–DCP would accumulate to this extent in field situations, there could have been serious ecological effects if the bacterial strain had been released into the environment, because mycorrhizal fungi are crucial to soil fertility and nutrient uptake by plants. This was the first report of an ecological effect induced through the accumulation of a metabolic by-product from a genetically modified microorganism intended for bioremediation of contaminated soil.

Bioremediation offers great benefits for decontamination of both industrial and agricultural land. The use of GMOs in bioremediation may bring major environmental benefits, but rigorous risk analysis must be carried out.

Impacts on soil ecology

Microcosms In the *Pseudomonas* study, the adverse ecological effects due to 2,4–D breakdown were observed in microcosms: controlled, reproducible, laboratory systems that attempt to simulate conditions in the real world. Microcosm studies play an important role in the ecological risk assessment of microbial GMOs prior to their release into the environment. In these studies, microcosms with GMOs are compared to those with equivalent unmodified organisms. A range of ecological effects or end points can be looked for in microcosms, 27 of which have been listed by Bailey *et al.* (1994). These include changes in metabolic (respiratory/carbon dioxide) activity among soil organisms, the total number of viable bacteria at selected time intervals, effects on species diversity, effects on plant roots and symbiotic fungi, and resource utilization.

Soil was collected from the field in the *Pseudomonas* study and treated so that the microorganisms in it were preserved. The effects on soil fungi were quantified by measuring the extent to which they spread in microcosms containing the GMO, compared to microcosms containing wild-type *Pseudomonas*. The toxic effects on the fungi were observed only in the microcosms containing the modified *Pseudomonas*. The effects were not immediate, so the study had to be conducted over several weeks.

Breakdown of plant residues   Another commonly proposed application of genetically modified soil microorganisms is the breakdown of unwanted plant material, such as crop residues. These GMO releases would be highly beneficial in situations where the burning or ploughing-in of plant material is undesirable, for example, in fields where the burning of crop residues can cause air pollution and smoke that is hazardous to road users.

Donald Crawford and colleagues at the University of Idaho, USA, have studied the decomposition of plants by genetically engineering the soil bacterium *Streptomyces lividans*. The modified bacteria had an enhanced capability to degrade lignin, which is particularly abundant in woody plants and trees. The modified *S. lividans* degraded lignin more quickly by virtue of an inserted transgene that produced an enzyme called lignin peroxidase. This enzyme was present at levels three to four times higher than in unmodified *S. lividans*. Only a relatively narrow range of microorganisms degrade lignin, in comparison to other plant-derived organic molecules, so modified bacteria that fulfilled this role would be particularly desirable.

However, the modified *S. lividans* study was also the first report of a genetically engineered microorganism having a measurable effect on a biogeochemical cycle in the soil. The more rapid breakdown of lignin by the modified bacteria caused an increased evolution of carbon dioxide (CO_2), the end-point measured in microcosms, compared to wild-type bacteria. Wild-type bacteria in the vicinity could also utilize the extracellular enzyme produced by the modified bacteria, to break down lignin faster and produce further CO_2. In other words, a side effect of the release of the modified bacteria was that a short-term increase occurred in the turnover rate of lignin-derived organic carbon in the soil (Wang *et al.*, 1991). Changes in biogeochemical cycles can have knock-on effects within ecosystems, potentially affecting a range of other species.

Klebsiella are a group of soil bacteria that have been found on the roots of practically every plant examined for their presence. *Klebsiella* are effective at decomposing plant litter and are therefore ideal organisms to modify in order to dispose of unwanted plant material. In one study, *Klebsiella planticola* were modified to break down collected crop residues, while at the same time producing alcohol (ethanol), a saleable industrial product that would make the whole process economic. A gene from a mutated strain of bacteria was introduced into *K. planticola*, which does not normally produce alcohol, and this resulted in successful alcohol production accompanied by efficient breakdown of plant residues. It was proposed that the remaining sludge, which is rich in nitrogen, phosphorus and sulphur, be put back on to fields.

In a series of experiments conducted at Oregon State University, USA, agricultural soil containing a representative set of microorganisms, was placed into microcosms with wheat seedlings. Water alone was added to a third of the microcosms, unmodified *K. planticola* to another third, and the genetically modified *K. planticola* to the final third. After a week, all the wheat seedlings in the treatment with the genetically modified bacteria were dead, while the plants in the treatments with unmodified bacteria and water only were growing normally. The modified bacteria had attached themselves to plant roots in competition with strains naturally present in the agricultural soil, and had killed the plants by producing alcohol. Wheat plants are tolerant of 1 ppm (parts per million) alcohol, but the modified bacterium raised the level to around 17 ppm. The modified strain of *K. planticola* also increased in the soil the number of fungal-feeding nematodes, a type associated with depleted soil ecology. In a range of soil types, the fungi associated with plant roots were observed to decrease in the presence of the alcohol-producing *Klebsiella*. The symbiotic mycorrhizal fungal–plant-root relationship is vital for effective nutrient uptake by plants. The modified bacteria were shown to persist for long periods under conditions found in some soil ecosystems, and to stimulate changes in the soil biota that could affect nutrient cycling processes in the soil (Holmes *et al.*, 1999).

It is not known how well the modified *Klebsiella* would survive under field conditions. However, *Klebsiella* bacteria live on the roots of all plants. Therefore, if sludge containing alcohol-producing *K. planticola* had been spread over fields as a fertilizer, it could have been highly toxic to many plant species. The sludge could also have affected many other organisms via food chains, and have had a profound effect on the soil ecosystem (Ingham, 1998).

Laboratory experiments have therefore demonstrated how a range of ecological effects could be caused indirectly by microbial GMO releases, via breakdown products or through unpredictable effects on non-target organisms. It is noteworthy that normal agricultural soil was used in the Oregon *Klebsiella*, the Idaho *Streptomyces* (lignin breakdown), and the EPA *Pseudomonas* (2,4–D breakdown) studies. Agricultural soil typically contains around 600 million bacteria, approximately three miles of mycorrhizal fungal hyphae, about 10,000 protozoa, and between 20 and 30 beneficial nematodes, in a teaspoonful. It is a thriving ecological system. That the effects of transgenic *Streptomyces* on the carbon cycle and indigenous bacteria were identified was due solely to a variety of ecological end-points being tested on an agricultural-type soil. Elaine Ingham, the senior author of the Oregon *Klebsiella* study, has criticized tests routinely performed by the EPA to evaluate genetically engineered microorganisms for environmental release, in which they use microcosms containing sterile soil. Ingham has argued that if it's sterile, it's not really soil. The results cannot provide any information about how the GMO will behave in the field, in terms of effects on soil ecology or on other organisms (Ingham, 1998). In addition, no realistic data on the exchange of genetic information between different bacteria can be obtained in sterilized soils (Lorenz and Wackernagel, 1993).

Another application involving the degradation of plant residues is silage production. The type and number of microorganisms present directs the course of silage formation in silos. One of the important bacteria involved is *Lactobacillus plantarum*, which has been genetically engineered to enhance the formation of high-quality silage. In experiments with a transgenic strain of *L. plantarum*, it was found to decrease pH and become a more dominant microorganism in silos of ryegrass, compared to wild-type bacteria (Sharp *et al.*, 1992). A genetic modification therefore changed the structure of the microbial community.

In a microcosm study with five different strains of genetically engineered *Pseudomonas solanacearum*, community structure was found to be unique in each case. The numbers of protozoa – predators that consume *Pseudomonas* bacteria – were differentially affected by the genetic make-up of their prey (Austin *et al.*, 1990). Studies of this type underline the need to look at whole communities when assessing the potential ecological impacts of microbial GMOs.

Nitrogen-fixation The first report of a crop yield increase in the field as a result of a microbial GMO involved a modified strain of a nitrogen-fixing bacterium. Research Seeds of St. Joseph, Missouri, USA, developed the modified strain of *Rhizobium meliloti* (*Sinorhizobium meliloti*), which in September 1997 became the first transgenic microorganism to be approved for commercial release into the environment. *R. meliloti* naturally forms a symbiotic association with alfalfa. The bacteria live in root nodules and receive nutrients from the plant, while providing the plant with nitrogen 'fixed' from the air. The modified strain of *Rhizobium* was inoculated into alfalfa seed coatings, and designed to increase crop yields by enhancing nitrogen fixation. It was engineered with additional copies of two genes, one that regulates nitrogen-fixation (*nifA*) and one involved in energy transport within plants (*dctABD*). Antibiotic-resistance genes (streptomycin and spectinomycin) used as selectable markers, and a promoter from another bacterium (*Bradyrhizobium japonicum*), which fixes nitrogen in soyabean root nodules, were also included in the gene construct that was integrated into the modified strain of *R. meliloti*.

In field trials conducted in Wisconsin, USA, the modified *Rhizobium meliloti* strain increased alfalfa biomass by 12.9 per cent, compared with alfalfa seed inoculated with wild-type bacteria, and by 17.9 per cent compared to uninoculated control seed. However, biomass increases were observed only under conditions of low soil organic content, low soil nitrogen, and when there was limited competition with endogenous *Rhizobium* populations. Under a range of other soil conditions, no effects on biomass were observed. Therefore the potential yield benefits appear to be limited (Bosworth *et al.*, 1994).

The initial EPA evaluation assumed that the modified *Rhizobium* strain would behave like the wild-type bacterium in the field, which has been used commercially as an inoculant since 1895 without ill effect. However, the modified strain has several genetic differences to the wild-type, and the capability of changing the rate of uptake of a key biological element. The nitrogen cycle is one of the major biogeochemical cycles, and perturbations in it could have widespread ecological consequences. There is also the possibility that the modified strain could cross-inoculate other legumes, which might increase their ability to become weeds. No adverse environmental impacts have been recorded during small-scale trials, conducted since 1989, or during a larger three-year study, which confirmed that moderate alfalfa yield increases occurred under nitrogen-depleted conditions. However, the latter study also found that if the gene construct inserted itself at certain poorly characterized positions in the genomes of modified *Rhizobium*, then it could be detrimental to alfalfa yield and may have other unexpected consequences (Scupham *et al.*, 1996). More caution is required when assessing the impacts on soil ecology of microbial GMOs that can potentially alter biogeochemical cycles.

Baculovirus bioinsecticide

A baculovirus is a virus that specifically attacks insects. Solutions of baculoviruses have been used worldwide on crops as bioinsecticides against caterpillars of Lepidoptera (moths and butterflies). They degrade readily in the soil, are inert outside host cells, have a narrow host-range, and are therefore considered selective and safe. However, their use has been limited because they kill insect pests slowly, enabling caterpillars to continue feeding for up to several days before dying. Genetically modified baculovirus, containing genes for insect-specific toxins not normally found in the viruses, have been designed to be more efficient killers of caterpillars on cabbage and other vegetables. Modified baculoviruses kill insects 25–40 per cent faster than equivalent non-modified baculoviruses. This could represent significant yield loss reductions, achieved with minimal environmental impact (Moscardi, 1999).

Research groups in the USA and UK, developing and testing modified bioinsecticides, have assumed that they pose little ecological risk, based on previous experience with unmodified baculovirus. However, critics of the field-testing of genetically modified baculovirus have suggested that they might escape to infect the caterpillars of non-target moth and butterfly species. The insect-toxin transgenes might also cross into other types of virus, with unpredictable results. Field experiments in Oxford, England, carried out by the Institute of Virology and Environmental Microbiology (IVEM) during the 1990s, particularly alarmed ecologists because they were conducted close to Wytham Wood, an important ecological study area and nature reserve (Cory *et al.*, 1994). The IVEM team showed that for 58 moth and 17 butterfly species, a sub-sample of British Lepidoptera, no difference in host-range occurred between modified and wild-type baculovirus. Some native moth species were found to be susceptible, for example, the lime hawkmoth (*Mimas tiliae*) and the privet hawkmoth (*Sphynx ligustri*), but only at high doses (Bishop *et al.*, 1988). It was also stressed that the baculovirus was designed as an insecticide that disappears rapidly from the environment, and not as a biological control agent that becomes permanently established (Hammock, 1991). Nevertheless, the data suggested that around 5–10 per cent of British Lepidoptera were potential hosts, representing 125–250 species, including some of great conservation value (Williamson, 1991; Williamson, 1996).

No genetically engineered baculovirus had been approved for commercial release at the time of writing. Given the balance of ecological concerns and the benefits gained through quicker insect kills, their commercialization may not be worth the risk.

Genetic exchange

Most of the proposed environmental releases of genetically engineered microorganisms have concerned bacteria. Once established in the environment, microbial GMOs can exchange genetic information with other microorganisms via three principal mechanisms: conjugation, transduction, and transformation.

Conjugation   Genetic material can be directly exchanged when two bacterial cells come into contact, through a mechanism called conjugation. DNA is transferred in the form of circular structures called plasmids. It is the main mode of sexual reproduction in bacteria, as well as in algae, fungi and certain other microorganisms. Conjugation is dependent on the physical distance between bacterial colonies, and their encounter rate. It is more likely to occur between bacteria that are closely related, but it is not restricted to bacteria of the same species. Chemical signals passing between bacterial cells are thought to influence the frequency of conjugation (Davis, 1980).

Conjugation is the principle means of genetic exchange for soil bacteria, among which high frequencies of plasmid transfer have been recorded. The presence of actively growing plant roots can stimulate plasmid transfer (van Veen *et al.*, 1994). Conjugation is also common among bacteria in freshwater and marine environments. Marine bacteria congregate on surfaces to form biofilms, for example, where conjugation is favoured (O'Morchoe *et al.*, 1988; Zilinskas, 1998).

There are also examples of conjugative transfer of genetic information from bacteria to higher organisms. The soil bacterium *Agrobacterium tumefaciens* infects plants, causing crown gall disease, by passing its plasmids into the chromosomes of plant cells. The foreign genes direct the plant into making the galls, in which the bacterial colonies thrive. Genetic engineers have exploited this system, to get their own foreign genes-of-interest into plants, rather than the genes that cause the disease.

Transduction   The transfer of genetic material from one bacterium to another via another organism, usually a virus of bacteria called a bacteriophage, is called transduction. Transduction is known to occur frequently among soil bacteria, via the numerous viruses in soil ecosystems (Levy and Marshall, 1988). Free-living viruses are also relatively common in marine environments, and viral-mediated gene transfer has been demonstrated between different species of marine bacteria (Chiura, 1997).

Transformation   A bacterium can take up extra-cellular genetic material directly from its immediate environment by a mechanism called transformation. This DNA is present in a free form, called naked DNA. A bacterium is said to be in a state of competence when it readily accepts naked DNA, because specific genes are expressed at this time that aid in binding DNA and integrating it into the bacterial genome. Transformation is common among soil bacteria, and has been demonstrated in *Bacillus subtilis, Rhizobium melilotis,* and species of *Pseudomonas* and *Streptomyces* (Stewart, 1992; Lorenz and Wackernagel, 1994). Transformation is an important method of genetic exchange among bacteria in aquatic environments, due to the relatively high concentrations of free DNA in water (Lorenz and Wackernagel, 1993). Transformation may manifest itself as the gain of a new trait.

Antibiotic-resistance genes naturally occur on bacterial plasmids and are therefore commonly transferred between bacteria (Talbot *et al.*, 1980). Bacterial GMOs can transfer antibiotic-resistance transgenes to other microorganisms via conjugation. Microorganisms can also pick up any antibiotic-resistance transgenes in the soil around the roots of transgenic plants, where they are commonly used as selectable marker genes, via transformation. Therefore, taking all the routes of genetic exchange together, there is ample opportunity for transgenes to spread in the environment, even between 'non-interbreeding' species.

Risk containment and assessment

In the early days of genetic engineering, concerns were raised about the biosafety of bacteria containing recombinant DNA in the laboratory. These concerns led to the first conference on the biosafety of GMOs, held in Asilomar, California, in 1975. Scientists decided to practise self-control over their experiments, and a framework for the biological containment of GMOs was set out. Modified bacterial strains, commonly used in experiments aimed at understanding gene regulation, were subsequently engineered so that they could not survive outside the laboratory. This was achieved by making their growth dependent on compounds not readily available in nature, or 'crippling' them, for example with 'suicide genes' that ensured they had a limited lifespan. These systems were not 100 per cent effective. Mutations, for instance, could invalidate them. In hindsight, however, these containment measures were stringent and commendable. Techniques for the safe laboratory containment of bacterial GMOs have continued to advance. One estimate suggested that up to eight separate suicide-type mechanisms, acting independently, might be needed to ensure near 100 per cent safety. Microorganisms with multiple self-destruct systems are now being developed (Szafranski *et al.*, 1997; Zilinskas, 1998). However, the guidelines for containing GMOs have gradually been watered down over time as experience with them is gained.

Genetically engineered microorganisms in laboratories have a history of safe containment. They are now also commonly used in fermenters on an industrial scale, to produce enzymes in the manufacture of foodstuffs, therapeutic drugs, and other products of the new biotechnology. Their containment in industrial units is such that the chance of a GMO escaping to the wider environment is small. However, the deliberate release of genetically engineered microorganisms into the environment is another matter. They will, for instance, be difficult to detect and retrieve. A number of decontamination methods have been employed against microbial GMOs in the field, including burning, tilling, removal of soil and subsequent autoclaving, and the spraying of field sites with antibiotics. However, if they become established, it would be extremely difficult to eliminate them completely or to mitigate their effects in the event of unforeseen ecological impacts. A precautionary framework for the release of microbial GMOs into the environment, along the lines of the Asilomar academic conference, has not been put in place by the multinational corporations that now control biotechnology.

Mathematical models for predicting the behaviour of introduced microbial GMOs could be used to reduce the possibility of ecological risk (e.g. Lewis *et al.*, 1996; van Veen *et al.*, 1994). Physiological changes could also be made to reduce their viability. The ecological impact of a microbial GMO could be reduced using suicide or crippling genes, for instance, as is the case for experimental laboratory bacteria. This would limit their persistence in the environment. However, GMOs that have been debilitated may also have their useful characteristics attenuated and their benefits reduced. When a genetically modified baculovirus was 'crippled' by removing genes for certain surface proteins, for example, it was no longer effective as a bioinsecticide. On the other hand, in many field applications, including bioremediation, it would be beneficial for microorganisms to resist degradation, which potentially increases ecological risk.

To assess ecological risk, it is necessary to have data on rates of survival, establishment, multiplication, dispersal, and transfer of transgenes to endogenous microorganisms by conjugation or other routes. However, obtaining such data in the field is often difficult, which has limited ecologists' ability to predict potential environmental impacts. One review of the effects of microbial GMOs on ecological processes concluded that surprisingly few studies had been conducted on survival and genetic exchange in natural environments, particularly in the area of bioremediation, where many releases are proposed (Doyle *et al.*, 1995). There is insufficient knowledge of how microorganisms behave in the field, and predicting how microbial GMOs will behave is therefore problematic.

Applications of genetically engineered microorganisms may bring great environmental benefits, particularly in the area of bioremediation. However, there may be considerable ecological risk. Microorganisms readily disperse over long distances, reproduce rapidly, are highly resilient, and have a great ability to exchange genetic material. Modifications to bacteria can alter biogeochemical cycles, which could have widespread impacts throughout ecological systems. Everything connects in ecology, and this must constantly be borne in mind when evaluating GMO releases.

4 Transgenic crops

The main focus of this book will be on transgenic crops, because of all genetically modified organisms (GMOs) they have been released commercially on the largest scale. Most transgenic crops have been modified for resistance to herbicides. Many have also been made resistant to a range of pests and diseases. The benefits to growers can be considerable, but several environmental concerns have emerged. In the future, a wider range of characteristics will be engineered into transgenic crops, including those for industrial and medicinal products. This chapter examines transgenic crop applications and the environmental factors that might limit their long-term deployment.

Herbicide-resistance

Weeds compete with cultivated plants for vital resources, such as moisture, nutrients and light. Uncontrolled weed growth can result in large yield reductions, or even total crop losses in the case of plants such as sugar beet that are poor competitors with weeds. Despite advances in herbicide formulation, monetary losses due to weeds can still be high. This is often because applications of herbicides are constrained by the damage they can do to crops. Many important weeds are related to crops and they share similar responses to herbicides. Herbicide-resistant plants are designed to be tolerant of weedkillers that otherwise might harm them. By making crops resistant to herbicides, weed management can be made more efficient, more flexible, and cheaper for growers.

Herbicide-resistance is the characteristic most commonly engineered into transgenic crops. By 1998, for example, 32 per cent of the total soyabean (*Glycine max*) crop in the USA was transgenic and herbicide-resistant. Crop seed and herbicides are usually supplied by the same company and sold as a package, accompanied by a signed agreement between company and grower restricting what herbicides can be sprayed on the transgenic crop. Transgenic crops therefore increase demand and market share for particular brand-name herbicides, at a time when they are starting to come out of patent. The commercial benefits of combined seed and agrochemical sales explain why herbicide-resistance has dominated the first wave of transgenic crops.

Canola   Herbicide-resistance has been sought in crops for some time. It is considered highly desirable, for example, in oilseed rape (*Brassica napus*). Canola is a type of oilseed rape bred for low levels of both its main monosaturated fatty acid (erucic acid) and the pungent glucosinolates that are characteristic of brassicas. Canola is the main type of oilseed rape grown in Western Canada (the name 'Canola' has been registered by the Western Canadian Oilseed Crushers Association). The low levels of saturated fatty acids in canola oil make it particularly desirable for culinary use. Canola in Canada was made resistant to one herbicide (triazine) in the early 1980s using conventional plant breeding methods (Beversdorf *et al.*, 1990). These cultivars have been grown alongside transgenic herbicide-resistance ones in Western Canada. However, the incorporation of herbicide-resistance by conventional breeding methods has been slow. Genetic engineering, in contrast, has proved very successful in producing a range of crops resistant to different chemical groups of herbicides.

Herbicide specificity   Different herbicide groups have different modes of action, and so different transgenes are integrated into crops to confer resistance to different herbicide groups. Resistance is therefore obtained to specific herbicide groups only, and plants remain susceptible to other herbicides. The transgenes used have mainly been isolated from strains of soil bacteria, which have acquired mutations enabling them to break down herbicides. Most commercially released transgenic crops have been resistant to either glyphosate or glufosinate herbicides.

Glyphosate   Glyphosate is an organophosphorus compound with a broad-spectrum action. It is a post-emergent herbicide, meaning that it must be applied after weed seedlings have started to grow, but it is then active against all growth stages throughout the year. It can be used to control most of the major weed species found in crops. Glyphosate works by inhibiting an enzyme and thereby blocking the synthesis of amino acids, the building blocks of proteins. This causes cessation of plant growth, and eventually death (Grossmann and Atkinson, 1985).

Monsanto's glyphosate product Roundup® is the world's best-selling herbicide, and has generated much of the company's wealth over recent years (Mendelson, 1998). The range of glyphosate-resistant transgenic crops produced by Monsanto is sold as Roundup Ready™. They were developed using genes from strains of soil bacteria (*Pseudomonas* and

Klebsiella), which cause an overproduction of the enzyme suppressed by the herbicide. This counteracts the effect of glyphosate and plants can grow unharmed. Monsanto (now merged with Pharmacia and Upjohn) have released Roundup Ready lines of soyabean, maize (corn; *Zea mays*), oilseed rape (including canola), sugar beet (*Beta vulgaris*), tobacco (*Nicotiana*) and cotton (*Gossypium*) (Hinchee *et al.*, 1993).

Glufosinate The herbicide glufosinate (glufosinate-ammonium) is a chemical within the PPT (phosphinothricin) family of weedkillers. It has a broad-spectrum action, and acts to inhibit an enzyme that plants use to assimilate ammonia. The herbicide therefore prevents ammonia entering a biochemical pathway and thereby blocks the synthesis of a key amino acid. Meanwhile, ammonia builds up to toxic levels in the plant (Oxtoby and Hughes, 1990).

Hoechst introduced the glufosinate herbicide Basta® into the market in 1981. Hoechst became part of AgrEvo in 1994 (now Aventis). Resistance to Basta is obtained in transgenic plants using a gene from a soil bacterium (*Streptomyces*). This transgene expresses an enzyme that effectively detoxifies the herbicide by altering its chemical structure (De Block *et al.*, 1987). Glufosinate-resistant transgenic crops are marketed by Aventis under the name Liberty Link™.

Bromoxynil and others Commercial crops have also been made resistant to a range of other herbicides. Monsanto's BXN® cotton contains a transgene from a soil bacterium (*Klebsiella ozaenae*) that confers resistance to bromoxynil herbicides. The bromoxynil herbicide Buctril®, manufactured by Rhône–Poulenc (now Aventis), is used on transgenic cotton, where it fulfils a need for a broad-leaf herbicide that does not damage the crop.

DuPont have developed a range of sulphonylurea herbicides (Glean®, Synchronicity® and Release®), to which transgenic crops (STS® system) are being developed. Conventional breeding, with mutated plant lines, has led to the development of resistance to triazine and imidazolinone herbicides. Imidazolinone herbicides were originally developed by American Cyanamid and are sold under a range of brand-names (Contour®, Resolve®, Lightning®). IMIcorn™ is a variety of maize, for example, resistant to imidazolinone herbicides.

Case for The rapid take up of herbicide-resistant crops reflects grower satisfaction. Weed management is becoming more cost-effective, and economic gains are being made. In addition, herbicide-resistance can reduce the number of herbicide sprays required on a particular crop, and therefore have environmental benefits. Herbicide sprays can be used, for example, to kill all weeds after a crop has emerged, negating the need for an additional pre-emergence spray against those weeds that respond to herbicides in a similar way to the crop (Hinchee *et al.*, 1993). In some cases, the use of post-emergence broad-spectrum herbicides may also reduce the need for soil cultivation, which helps reduce erosion and conserves soil organisms and moisture. Soil conservation is advised for canola, for example, but it must be accompanied by good weed management, as the crop is initially slow growing and can easily be out-competed by weeds.

Case against   Rather than reducing herbicide use, herbicide-resistant crops may increase it. An upper limit on spraying exists with susceptible crops, above which damage occurs. In herbicide-resistant crops, however, there may be a tendency to overspray in order to maximize yields, as there will be no adverse consequences. Herbicide-resistance is opening up new applications and markets for weedkillers, so increasing the overall amounts of particular herbicides put into the environment. Companies have increased their production of herbicides, including glyphosate, to coincide with the release of transgenic crops. Meanwhile, industry has requested that higher levels of herbicide residues be allowed on transgenic crops. Permitted residue levels on soyabeans, for example, have been raised in the USA and Europe in recent years, and similar increases have been sought worldwide.

The ecological risks associated with herbicide-resistant crops will therefore in large part depend on whether they increase or decrease the amount of herbicide sprayed. Glyphosate and other herbicides used on transgenic crops are indiscriminate plant poisons, and increased spraying could impact on a wide range of herbaceous plants growing around farmland. Meanwhile, weed-free fields deprive wildlife of food resources, which can in turn reduce resources for creatures further up the food chain. The possible impact on biodiversity has been one of the major concerns relating to the proposed widespread cultivation of herbicide-resistant crops in Europe.

Much of the ecological concern expressed about herbicide-resistance is therefore not a consequence of genetic engineering *per se*, but the manner in which crops are grown in commercial situations. These concerns therefore also apply to conventionally bred herbicide-resistant crops. The herbicides to which transgenic crops have been made resistant have been chosen partly because they are less environmentally damaging than many other herbicides. However, claims that they do little ecological damage need to be examined critically. If more of these herbicides enter the environment as a result of the cultivation of herbicide-resistant crops, then adverse ecological impacts are likely.

Glyphosate and the environment Glyphosate herbicides have been used for around 25 years, and are generally regarded as among the safest weedkillers available. They are used globally, in agricultural, forestry, aquatic and garden situations, and sold under a range of brand-names, including Rodeo®, Accord®, and Vision®. Monsanto claims that during normal use, its glyphosate herbicide Roundup is practically non-toxic to birds, fish, aquatic invertebrates, algae, honeybees, other beneficial arthropods, and earthworms. The company has marketed Roundup as an environmentally friendly product for many years, but there is evidence of toxicity to non-target organisms (Cox, 1995a).

A number of studies have shown glyphosate herbicides to be acutely toxic to fish, especially salmonids, including rainbow trout, sockeye salmon, and coho salmon (Folmar et al., 1979; Holtby and Baillie, 1987; Liong et al., 1988; Mitchell et al., 1987; Servizi et al., 1987; Wan et al., 1989). In addition to direct toxicity, run-off from farmland or drift from aerial spraying can stimulate undesirable eutrophication (nutrient enrichment) in waterways (Austin et al., 1991). This can affect species composition within communities, for example, with the emergence of a dominant species and reduced biodiversity, which can have consequences throughout an ecosystem. Glyphosate herbicide has been shown to reduce midge populations, for instance, which are an important component of aquatic food chains (Buhl and Faerber, 1989).

In a study of the effects of 62 pesticides on beneficial organisms, glyphosate was tested on 19 species and found to be moderately harmful to a predatory beetle (*Bembidion*), and slightly harmful to a parasitic wasp (*Trichogramma*), a predatory mite (*Typhlodromus pyri*), a ladybird (*Semiadalia*) and a lacewing (*Chrysoperla carnea*) (Hassan et al., 1988). Most beneficial organisms, including honeybees, appear not to be harmed by concentrations of glyphosate typically encountered in the field. A common New Zealand earthworm showed retarded growth and development in response to relatively low concentrations of glyphosate in a laboratory study (Springett and Gray, 1992). However, this contrasts with other research on herbicide toxicity and earthworms, while glyphosate is ranked as being relatively non-toxic in a standard text on earthworm biology (Edwards and Bohlen, 1996; Monsanto, 2000a).

The broad-spectrum action of glyphosate means that it is toxic to a wide range of plants, including many vulnerable and rare species that

grow in the vicinity of agricultural land. The US Fish and Wildlife Service
has identified 74 endangered plant species it believes are at risk as a result
of glyphosate use (EPA, 1996). Glyphosate is relatively non-volatile and
does not drift laterally from fields as a vapour when sprayed at ground
level, as many other pesticides do. However, when sprayed from aircraft
or helicopters it can drift some 400–800 metres downwind of fields into
streams and neighbouring habitats (Cox, 1995b). In the mid-western USA,
the cultivation of herbicide-resistant cotton and soyabean has been
accompanied by an increase in the aerial spraying of herbicides. Maize
growers downwind of such spraying have reported crop losses as a result
(Lappé and Bailey, 1999). This puts pressure on farmers in some areas to
conform with their neighbours and grow Roundup Ready crops that will
not be damaged by clouds of descending herbicide. If the trend towards
aerial spraying continues, native flora will be at increasing risk from the
herbicide.

Like that of all herbicides, glyphosate persistence in the field depends
on soil type and temperature. Glyphosate binds to soil particles and under
certain circumstances can remain active for many weeks. The fate of
radioactively labelled glyphosate can be followed in the soil to determine
its persistence. This has been recorded as varying from 3 to 141 days (Cox,
1995b; Mendelson, 1998). One report claimed that residues remained
active on vegetables for up to a year after treatment, but this activity was
most likely to have been due to individual radioactive carbon molecules
(C^{14}) from degraded glyphosate and not the weedkiller itself (Monsanto,
2000a).

Glyphosate herbicide will frequently be active in the soil long enough
to inhibit the growth of the soil fungi that enter into important symbiotic
relationships with plant roots. Toxic effects on mycorrhizal fungi in soil
have been recorded in a number of studies (Chakravarty and Chatarpaul,
1990; Cox, 1995a; Estok *et al.*, 1989). Some soil fungi, such as *Penicillium
funiculosum*, are more susceptible than others to the toxic effects of glyphosate, leading to probable changes in community structure and possible
implications for soil fertility generally (Abdel-Mallek *et al.*, 1994). If mycorrhizal fungi in the soil were to be affected, owing to excessive spraying or
particular soil conditions, it could have profound effects on a plant's ability
to take up resources.

Cases of death have been attributed to glyphosate herbicides, although
these have mainly been farmer suicides where relatively large quantities
have been drunk. In normal use, it should not be acutely toxic (Monsanto,
2000a). However, a Californian study showed that glyphosate was the
third most frequent cause of illness amongst agricultural workers (Cox,

1995b). Most reported health incidents involving glyphosate have involved minor eye irritations (Monsanto, 2000a). Nevertheless, mammalian studies suggest that glyphosate herbicides may be carcinogenic and could disrupt reproduction (Cox, 1995a). Toxicity tests should be based on entire formulations and not just active ingredients. The main toxicity of glyphosate herbicides, to fish, mammals and humans, is not from the active ingredient (glyphosate), which has a relatively low toxicity, but from the 'inert' ingredients in the formulation. A surfactant (called POEA), a chemical that helps the active ingredient to pass into the plant, for example, is the ingredient mainly responsible for human deaths when the herbicide is swallowed (Cox, 1995a; Martinez and Brown, 1991; Sawada *et al.*, 1988).

Glufosinate and the environment   Glufosinate is also considered to be a relatively safe pesticide. It has a variety of uses as a broad-spectrum herbicide, and is favoured in situations where vegetation needs to be completely cleared. It is also used as a pre-harvest treatment in crops such as oilseed rape, to kill foliage so that harvesting is made easier. Its persistence in the soil varies with environmental conditions, with a half-life of up to twenty days, according to a standard pesticides manual (Tomlins, 1997), although it can persist for over 60 days in some cases (Cox, 1996).

Glufosinate is highly soluble and can contaminate groundwater and run-off from fields. If it gets into streams and rivers it can be highly toxic to aquatic life, including invertebrates. Its broad-spectrum action means that it can seriously reduce plant biodiversity around agricultural land. Changes to the flora of field margins have been reported one year after spraying with glufosinate, in comparison with an untreated control area. Rare or scarce arable flower species declined at the expense of the more dominant and aggressive weed species such as black grass (*Alopecurus myosuroides*) and cleavers (*Galium aparine*) (Genewatch, 1999).

Bromoxynil and the environment   More herbicide is sprayed on cotton than any other crop, and regulations on spraying have in the past been lax. There were effectively no restrictions on the spraying of bromoxynil (3,5-dibromo-4-hydroxybenzonitrile) on transgenic bromoxynil-resistant cotton during the first few years of its commercial cultivation in the USA. Tolerance levels for the herbicide in cottonseed were such that high levels of spraying could occur. Many concerned groups and individuals argued that the EPA had made a mistake with this approval for the following reasons. Bromoxynil is a toxic chemical. It is a possible human carcinogen and causes birth defects in mammals. Although cotton is primarily a non-food crop, cottonseed oil is used in food products and crushed seed is used in animal feed. In addition to the toxicity of the active ingredient, transgenic bromoxynil-resistant cotton degrades bromoxynil into a metabolite called DBHA (3,5–dibromo–hydrobenzoic acid), which forms a major part of the herbicide residue (about 80 per cent) on transgenic cotton. This residue is not found on conventional cotton. Little is known about the toxicity of DBHA and the EPA did not initially set any tolerance limits for it. Exposure to this residue could potentially result in the poisoning of animals or humans (Lappé and Bailey, 1999).

Bromoxynil-resistant cotton was grown on 170,000 hectares in 1997 and it was to be extended to around 10 per cent of the total US cotton crop in 1998. However, the EPA added further safety factors to all pesticide tolerances in 1998. They refused to make a special case for bromoxynil in cottonseed. This effectively meant that bromoxynil could no longer be sprayed on cotton, because the residue levels would be too high. Rhône–Poulenc (Aventis) and Stoneville Pedigree Seed of Memphis, which markets bromoxynil-resistant cotton, appealed against this decision, which severely limits the usefulness of this transgenic crop in the USA (Holmes, 1998).

Weed resistance and weed shifts   When a single herbicide is repeatedly sprayed, as will be the case for transgenic crops where seed and herbicide are marketed together, the chances of herbicide-resistance developing in weeds increases. The selection pressure on weeds is much greater than if different herbicide groups were used in rotation. Many cases of weed resistance to herbicides are known, and the cultivation of transgenic crops risks creating further cases. Continual spraying of one herbicide can also lead to a 'weed shift', towards a community of weed species that are more tolerant of a particular herbicide. In the long term this can reduce a herbicide's effectiveness. In Roundup Ready cotton, increases in spurges, bermudagrass, barnyard grass and other weeds that are more tolerant of glyphosate has occurred, while in BXN cotton a weed shift has favoured nutsedge, sicklepod, pigweed and other weeds that are relatively tolerant of Buctril (Anon., 2000a). These scenarios would complicate weed management in the long term, rather than aid it.

Herbicide-resistant crop plants might themselves become weeds in other crops. If they grew in the following year's crop on the same site, as 'volunteer' weeds, they may resist herbicides sprayed on to those crops. The pollen from herbicide-resistant crops could also fertilize wild relatives, resulting in hybrid weeds with herbicide-resistance. The imminent introduction of commercial herbicide-resistant wheat, for example, could result in grassy weeds, such as goat grass, becoming herbicide-resistant by cross-fertilization. The potential for transgenic crops to become invasive, or to transfer transgene to wild relatives, is the subject of the next two chapters.

Crop performance Although farmers have been enthusiastic in their take-up of transgenic crops, some of the benefits derived from them have been called into question. A study by the United States Department of Agriculture's (USDA) Economic Research Service showed that, in 1998, yields were not significantly different between transgenic and non-transgenic crops in 12 out of 18 crop/region combinations. In the case of herbicide-resistant cotton, maize, and soyabean, farmers frequently used similar amounts of herbicide as on conventional crops. Therefore these farmers were making no savings on herbicide costs. Roundup Ready cotton grown in the Mississippi delta in 1997 (amounting to about 0.6 per cent of the transgenic US cotton crop) had very poor yields owing to the phytotoxic effects of Roundup, from which it should have been protected. This demonstrated that the performance of the transgenic cotton could be erratic and subject to climatic conditions (cool weather in this case). Data from Arkansas has shown that Roundup Ready soyabeans did not out-perform the best non-transgenic varieties in terms of yield. Overall, when yield increases have occurred with herbicide-resistant crops, they have been in the order of 5–30 per cent (USDA, 1999).

The stems of Monsanto's herbicide-resistant soyabeans have been found to split in hot climates. This resulted in crop losses of up to 40 per cent during a hot spring in Georgia, USA. This was a major setback for the company, which had hoped to market this transgenic crop in Latin America. The area under soyabean, for example, has been increasing at a rapid rate in Brazil. In laboratory experiments, transgenic soyabeans became stunted and more brittle at high temperatures, unlike conventional ones, due to a change in plant physiology. The enzyme that conferred resistance to glyphosate in soyabeans also affected a metabolic pathway in the plant, resulting in lignin overproduction. The increased levels of lignin caused the brittle stems that are more easily split (Coghlan, 1999). This unpredicted effect on a major biochemical pathway was noted only after the commercial release of this transgenic crop.

Herbicide-resistance has proved of great benefit to many growers, but in a number of crop situations the benefits have not been great. In addition, consumers do not benefit from the fact that crops are herbicide-resistant. Herbicide-resistance may lead farmers to spray greater amounts of herbicide. If this happens it may be detrimental to the environment, having particularly adverse effects on biodiversity.

Insect-resistance

Pest-resistance is the second most common trait engineered into transgenic crops. All commercially released transgenic crops resistant to insect pests have incorporated genes derived from just one soil bacterium: *Bacillus thuringiensis* (*Bt*). This bacterium naturally accumulates large amounts of toxic protein during its spore or resting stage. If insects ingest these spores, the toxin causes their guts to become paralysed, feeding stops and death ensues. The toxins kill only larval insect stages and are not toxic to adults or other organisms. *Bt* spray formulations have been used as bioinsecticides since the late 1950s, and are an important pest-control tool for organic growers. The use of *Bt* sprays, however, has been constrained by high production costs and poor persistence in the field. Transgenic *Bt* crops were envisaged as fulfilling an important role within Integrated Pest Management (IPM) programmes, in which pests could be controlled with less reliance on synthetic insecticides (Peferoen, 1997).

Thousands of different strains of *B. thuringiensis* are now known. The first was discovered in 1911 in Thuringia, Germany; hence the name. Toxicity is to specific insect groups: for example, the strain *B. thuringiensis kurstaki* kills certain species of Lepidoptera (moths and butterflies) and *B. thuringiensis tenebrionis* kills certain Coleoptera (beetles). The toxin genes from several different *Bt* strains have been integrated into a wide range of transgenic plants. In commercial crops, the *Cry1A* gene has been used to confer resistance to lepidopteran pests, such as the European corn borer (*Ostrinia nubilalis*), while the *Cry3A* gene has been used to confer resistance to coleopteran pests, including the Colorado potato beetle (*Leptinotarsa decemlineata*) (Barton and Miller, 1993).

Over forty different types of transgene, in addition to *Bt* genes, have been used to confer insect-resistance in experimental programmes, including genes encoding for proteinase (protease) inhibitors and lectins (Schuler *et al.*, 1998). These provide a general anti-herbivore defence, in contrast to specific *Bt* toxins, and therefore could potentially confer resistance to a range of insect pests (Hilder *et al.*, 1990). Proteinase inhibitors, widely found in seeds and storage organs, form complexes with insect digestive enzymes, preventing them from breaking down proteins. A number of distinct families of proteinase inhibitors have been identified, and over fourteen different plant proteinase-inhibitor genes have been introduced into crop plants. The trypsin inhibitor from cowpea (*Vigna unguiculata*) is the most active of these.

Lectins, found in the seeds and tissues of legumes and other plants, are proteins which bind to carbohydrates. Some of them are toxic to insects, particularly species not adapted to plants in the pea family (Leguminosae). A lectin gene from the garden pea (*Pisum sativum*), for example, has been transferred into potato (*Solanum tuberosum*) to confer resistance to Colorado potato beetle and other pests. Lectin genes have also been identified from snowdrops (*Galanthus nivalis*). The snowdrop lectin agglutin acts as an antifeedant against aphids, and has also been incorporated into potatoes. No crop modified with proteinase inhibitors or lectins had been released commercially by the end of 2001.

***Bt* crops** Commercial insect-resistance crops are currently limited therefore to *Bt* crops. Monsanto has been the main player, with its patented lines of Bollgard® cotton, Yieldgard® maize, and NewLeaf® potato. The first *Bt* crop (cotton) was grown commercially in 1996. By 1999, *Bt* cotton and *Bt* maize accounted for about a quarter of the total USA cotton and maize crops, respectively. However, controversy has surrounded the introduction of *Bt* maize into Europe, where several countries have been opposed to its importation and cultivation (Nottingham, 1998).

Case for The reduction in insecticide spraying on *Bt* crops brings economic benefits to farmers. Insect-resistant crops can also be beneficial to the environment. They require less insecticide, because the insecticidal material is contained in the plant tissue itself. This favours biological control, because natural enemies are less impacted by spraying; while other beneficial and non-target organisms, such as honeybees, have a reduced exposure to insecticides. Spray drift and its associated problems, such as groundwater contamination, are also reduced. Pest control is potentially more efficient, because the control agent is continuously present in the field and expressed in plant parts that are often difficult to reach by spraying. Human health benefits include fewer farm-workers being poisoned.

Case against It should be noted that *Bt* confers resistance only to certain insects. Therefore, insecticide sprays still need to be applied against other pest species. The USDA Economic Research Service survey showed that during 1997, seven out of twelve sites showed no statistical difference in the overall amounts of insecticide sprayed on transgenic as compared to non-transgenic cotton (USDA, 1999). The deployment of insect-resistant crops has raised concerns that pests might rapidly evolve resistance to insecticidal toxins that are continuously expressed in plants. Therefore, complex resistance management programmes will be needed if insect-resistant crops are to be of more than short-term benefit.

Resistance management   The cultivation of large areas of *Bt* crops could result in increased resistance in insects to *Bt* in general, including the *Bt* sprays that are important bioinsecticides used by organic farmers. Resistance to *Bt* has already occurred in some pest insects (Tabashnik, 1994; Tabashnik *et al.*, 1997). The strategy adopted for delaying the development of insect resistance to transgenic *Bt* plants is to use high-dose transgenic cultivars and establish refuges of plants that do not produce *Bt* toxins. High-dose transgenic plants cause 100 per cent insect mortality in laboratory studies. In field situations this is desirable to prevent selection for resistance genes, which will occur if resistant individuals survive and breed.

Meanwhile, pest populations that are susceptible to *Bt* toxins are maintained in refuges. This acts to dilute the overall effect of resistance genes, because susceptible insects from the refuges mate with resistant ones. However, there is disagreement among scientists about the appropriate refuge size for keeping resistance in check. The refuge sizes that were initially suggested, in the order of four per cent of the transgenic crop, now appear completely inadequate. Monsanto has recommended refuges of twenty per cent non-transgenic crop for Bollgard cotton, although insecticides can be used on this non-transgenic crop. However, some researchers have suggested that no insecticides should be sprayed on refuges of this size, with the risk that they might sustain heavy yield losses, while a patchwork of 40 per cent non-transgenic and 60 per cent transgenic might be required to delay resistance development if insecticides are used (Feldman and Stone, 1997; Gould, 1998).

The high-dose/refuge strategy will work only if two assumptions are correct: a) that inheritance of resistance is recessive (insects will be resistant only when copies of the resistance gene are received from both parents), and b) that random mating occurs between susceptible and resistant insects. Both these assumptions have recently been challenged. One study revealed that resistance to the *Bt* toxin gene appears to be inherited in an incompletely dominant (rather than recessive) manner in European corn borer, a pest against which *Bt* maize is widely deployed (Huang *et al.*, 1999). Random mating, meanwhile, might not occur if transgenic crops affect insect behaviour or physiology in ways subtly different from conventional crops. When pink bollworm (*Pectinophora gossypiella*) caterpillars were fed on *Bt* cotton in a laboratory study, for example, the timing of their development was altered. Laboratory strains of pink bollworm resistant to *Bt* toxin took longer to develop on transgenic plants than non-resistant strains raised on unmodified cotton. Such asynchronous development favours non-random mating (Liu *et al.*, 1999). The effectiveness of the refuge strategy will be reduced if resistance is not always a recessive trait or if random matings do not occur between resistant and susceptible insects. The testing of these basic assumptions should have been done before the extensive planting of *Bt* maize.

In practice, refuge strategies will need very careful monitoring if resistance development is to be avoided. However, in a survey conducted in 2000, about a third of US farmers failed to follow the rules governing refuges when planting *Bt* maize resistant to European corn borer (Anon., 2001a).

Crop performance   Yields have generally been higher in Bollgard *Bt* cotton than non-transgenic cotton, although not as high as predicted; and some serious cultivation problems have arisen. In Texas in 1996, for example, cotton was heavily attacked by the pink bollworm, an insect pest that it was engineered to resist. Feeding damage occurred on 800,000 hectares, although catastrophic losses were avoided by spraying with conventional insecticides. In this case, significantly more insecticide was sprayed on transgenic than on non-transgenic cotton (USDA, 1999). The Bollgard cotton in Texas produced too little toxin in some of its leaves for them to be injurious to the caterpillars, which preferentially fed on those leaves to avoid harm. This behavioural resistance raises questions about the effectiveness of the refuge strategy, as many insects within the *Bt* crop may not be killed, and about the possibility of insects circumventing insect-resistance in unexpected ways.

Transgenic *Bt* crops may have a range of other environmental impacts. Pollen containing *Bt* toxin can be toxic to non-target insects, such as honeybees and butterflies, while decaying plants can release *Bt* toxins into the soil, where they could be harmful to soil microorganisms. The potential ecological risks due to *Bt* crops will be discussed further in Chapter 7.

Virus-resistance

Disease resistance is the third most common trait engineered into transgenic crops. Many plant diseases have a viral origin, and these are responsible for high levels of crop loss. Virus-resistant transgenic plants have integrated into their genomes, sequences of viral nucleic acid, which interfere with the replication of any invading virus. This effectively immunises plants against viral attack (Tepfer, 1993). A transgenic sweet potato (*Ipomoea batatas*), for example, jointly developed by Kenyan researchers and Monsanto, confers resistance to feathery mottle virus

(FMV), a viral disease that can cause yield losses of around 50 to 80 per cent in some African countries. Virus-resistant crops can provide great economic benefits and safeguard important food supplies in virus-prevalent areas.

However, a unique set of ecological risks is associated with virus-resistant crops. In particular, concerns have been expressed about whether gene flow from transgenic plants to viruses could accelerate the rate of evolution of natural virus populations. This could lead to the emergence of novel disease-causing (pathogenic) viruses. This type of genetic exchange is a normal feature of virus evolution, and most viruses comprise a mosaic of differently rearranged genetic modules of diverse origin (Robinson, 1996).

Viruses have been shown to pick up and incorporate genes from transgenic plants. This was first demonstrated in the early 1990s, when an incomplete Cauliflower Mosaic Virus (CaMV) was injected into transgenic oilseed rape plants, engineered with the missing viral genes. The defective virus was able to reacquire the viral genes from the plant to become a fully functioning invasive virus, which caused normal CaMV symptoms in several experimental plants (Gal *et al.*, 1992). In another study, transgenic plants were infected with a strain of Cucumber Mosaic Virus (CMV) lacking a gene for a key protein that enables viruses to move between plant cells. The virus was able to reacquire this gene if it was engineered into the plant.

A similar recombination event was shown to occur with transgenic virus-resistant cowpea plants and cowpea chlorotic mosaic virus (CCMV). The virus was able to take up genetic material from the transgenic cowpea, although in this case it caused different symptoms to appear in plants compared to wild-type virus (Allison *et al.*, 1997). In another study, the genetic material of two viruses, East African Cassava Mosaic Virus (EACMV) and African Cassava Mosaic Virus (ACMV), which attack cassava (*Manihot esculenta*) in Uganda, was discovered to have recombined to form a novel and more virulent hybrid strain. This hybrid strain caused severe crop losses. It is suspected that the two viruses recombined within a cassava plant (Zhou *et al.*, 1997). The experimental evidence suggests that recombination between virus and plants, and between different viruses within plants, may not be unusual.

Virus-resistant transgenic crops contain viral genes in all their cells for the lifetime of the plants. This genetic material is therefore very accessible to plant viruses. Given the ability of viruses to acquire, recombine and swap genetic material, the deployment of large areas of viral-resistant transgenic crops may create the ideal conditions for new disease-causing viruses to evolve. A growing concern that new and virulent hybrid viruses could be produced led the USDA in August 1997 to outline proposed restrictions on transgenic crops engineered with genetic material from viruses. These restrictions included limiting the length of genetic sequences integrated into plants and banning the use of particular genes in virus-resistant crops (Kleiner, 1997a). Measures that could be used to minimize the ecological risks associated with virus-resistant transgenic crops include using only transgenes from viruses that are already common in particular crops. Possible recombinants, between viral transgenes in plants and commonly infecting viruses of those plants, should also be constructed and tested for pathogenicity, virulence and host range (Teycheney and Tepfer, 1999).

Environmental concerns also arise from the effect of the transgene on the plant itself. Resistance to CMV can be obtained in transgenic crops using a gene sequence from satellite RNA, a structure in the virus that can reduce viral symptoms. However, under some circumstances satellite RNA can exacerbate rather than attenuate viral symptoms. This was observed in Italy in 1996, when a naturally mutated satellite RNA sequence led to an epidemic of lethal tomato necrosis, which resulted in severe crop losses. Deleterious forms of satellite RNA were thought to be rare in nature, and therefore unlikely to occur in transgenic plants. But research has shown that deleterious satellite RNA can arise by mutation more commonly than previously thought; and that it has a selective advantage over the parent satellite RNA (Tepfer and Jacquemond, 1996). Therefore, such mutations could occur within transgenic plants, leaving crops more vulnerable to virus attack.

Virus-resistant crops may bring important benefits in terms of crop yield increases, particularly in the developing world. But their release into the environment is accompanied by significant ecological risks. In the USA, concern about the environmental impacts of virus-resistant plants may stall commercialization, although unrestricted commercial use is likely to continue in China and elsewhere.

The next generation

The transgenic crops released to date have been referred to as 'first generation' crops. They have been modified for a limited set of agronomic benefits. A 'second generation' of transgenic crops is being developed. They will have a much wider range of modifications and offer potentially greater benefits to society at large. The biotechnology industry has been particularly keen on the differentiation of transgenic crops in terms of generations. It distances them from miscalculations in the past, and suggests that things are moving forward. Thinking in terms of generations does help to put developments into a historical perspective, but it masks what is essentially a continuum. More complicated gene constructs will be used to achieve more ambitious modifications, but the technology and commercial motivations remain the same.

There is not enough room here to do justice to the range of transgenic crops currently being developed. Modifications, however, can be grouped into four main areas. First, increasingly complex modifications for agronomic benefits may result in crops that are resistant to a range of stress conditions (e.g. drought), and plants with enhanced nitrogen-fixing or photosynthetic abilities. Second, the cultivation of transgenic crops for industrial products will expand into new areas, such as bioplastics. Third, crops will be modified to produce functional foods, with enhanced nutritional qualities or extra vitamins. This will overlap with the fourth area, involving modifications for medicinal purposes, with transgenic crops producing pharmaceutical proteins and vaccines. The benefits of this technology could therefore be enormous. However, unique ecological risks must be considered in each of these areas.

Resistance to stress   Desertification is a growing problem, particularly in the developing world, and there is a need for crops that can grow on marginal land. Transgenic plants that tolerate drought conditions and high salinity have been engineered using genes from yeasts. A transgene expressing trehalose, a substance that enables baking yeast to exist in a dried state, can help plants to survive periods of drought; while a transgene expressing a protein that efficiently pumps sodium out of cells, utilized by yeasts that naturally thrive in saline environments, confers salt-tolerance on plants. Transgenes expressing trehalose and other proteins also confer a degree of frost-resistance. Experimental lines have been produced, but no commercial crops are yet available that tolerate environmental stresses. Such crops could provide great environmental benefits. Valuable water resources could be conserved not just in areas undergoing desertification, but also in all areas where crops need to be irrigated. Soil erosion could be stemmed in marginal agricultural areas, while stress-tolerant varieties could be grown over extended ranges. However, care must be taken because plants that withstand environmental stress can thrive in areas beyond their home range, and possibly become invasive.

Photosynthesis and nitrogen-fixation   Transgenic crops with more efficient means of photosynthesis or nitrogen-fixation could increase crop yields substantially. It would probably represent the single biggest advance since the major yield increases of the Green Revolution. However, in common with many next-generation modifications, solutions involving a single transgene and one foreign protein may be insufficient. Whereas adding one transgene accompanied by a promoter can confer herbicide-resistance, obtaining significantly enhanced efficiency of photosynthesis or nitrogen-fixation may require a number of transgenes. Key enzymes in the complex photosynthetic biochemical pathway have been identified and the genes responsible have been targeted first. The enzyme ribulose biphosphate carboxylase (rubisco), for example, brings carbon dioxide into the plant's metabolic cycle for the first time. Increasing the efficiency of this stage through genetic engineering could help to produce yield increases. However, changes in one step of a biochemical cycle may have effects in a number of different areas.

Experimental transgenic maize, developed by David Lighfoot and his colleagues at the Southern Illinois University at Carbondale, USA, has been engineered to take up nitrogen more efficiently from the soil. It contained a gene from a bacterium (*Escherichia coli*) that expresses the enzyme glutamate hydrogenase. The maize plants had higher protein content, increased size, and a greater tolerance of soil ammonia than unmodified plants. The roots and leaves contained double the amount of amino acids, the building blocks of proteins. This could translate into yield increases in the order of 5–10 per cent in the field, without the need for fertilizer applications. However, an unexpected doubling of sugar levels also occurred in the tissues of these plants. This would make the plants more attractive to insect pests, and might have destabilising effects on soil ecology (Dado and Lightfoot, 1997; Coghlan, 2000a). Crops with enhanced nitrogen-fixing properties could bring substantial economic, social and environmental benefits. However, nitrogen is an important element that cycles through ecosystems and the potential ecological risks of such crops would need to be carefully assessed.

Industrial products   Crop plants have been grown for industrial oils and other products for many years. However, genetic engineering has dramatically increased the scope for industrial modifications. Transgenic canola, for example, has been engineered to produce oil with high levels of oleic acids, such as erucic acid, for industrial use. Modifications to cotton may lead to the fibres being dyed on the plants, using transgenes expressing pigment, or having enhanced quality. These modifications could have environmental benefits in terms of reductions in the use of dyeing and processing chemicals.

Plastics were first expressed in transgenic plants in the early 1990s, using a transgene from a bacterium that makes a type of natural plastic in its cells (Poirier *et al.*, 1992). Maize and oilseed rape have been modified to produce these polyhydroxybutyrate plastics, which are biodegradable, unlike petrochemical-based plastics, which decompose slowly. But yields have been low, while polyhydroxybutyrates are relatively brittle and have limited use in industry. In 1999, Monsanto improved the original technique by producing commercial-grade PHBV (poly[3–hydroxybu-tyrate–co–3–hydroxyvalerate]) within plants in two building blocks, using four bacterial transgenes. Canola plants were engineered to express PHBV in their seeds. This raised yields, but they were still below levels consid-ered to be economically viable. Biopol (biodegradable natural plastic polymers) plants have been put on hold by Monsanto, although the company produces Biopol commercially in genetically engineered bacteria by fermentation (used by Greenpeace as a biodegradable alternative to PVC in their credit cards). However, Monsanto still hope to license plant Biopol technology in the future (Poirier, 1999). Plastic plants are consid-ered as a 'green' technology because they are produced from a renewable resource and are biodegradable. But there are 'hidden' environmental costs; for example, fossil fuels are still required to power the process that extracts plastic from plants (Gerngross and Slater, 2000).

Monsanto has envisaged plants, particularly canola, as biological factories, which in the future will produce oils, plastics and animal feed all in one crop. However, concerns about genetic contamination may dampen such optimistic visions. Crops for human consumption have been grown in the vicinity of crops grown for industrial purposes for many years. Transgenic cultivars of the same crop types grown for a wide range of industrial or medicinal purposes increase the risk of genetic contami-nation and the presence of undesirable products in animal feed or human food. This might emerge as the biggest environmental risk associated with transgenic crops expressing potentially toxic industrial products.

Functional foods   Transgenic crops have been experimentally modified for changes to their starch, oil, protein, sugar and vitamin content. Foodstuffs from these crops are likely to be marketed as 'functional foods' on the strength of their health benefits. A commercial line of transgenic high-starch potato can be used to produce French fries (chips) that take up less oil when frying. Transgenic canola has been modified to manufacture oils and margarines with ratios of saturated to unsaturated fatty acids that are also beneficial to health. The protein content of certain grains and legumes has been enhanced by adding amino acids that are traditionally absent, while certain proteins that are allergens have been removed from rice (*Oryza sativa*). Sugars in fruits and vegetables are being either enhanced or reduced to improve palatability. Nutraceuticals are foods with added nutritional and medical benefits. This is a potentially massive growth area, where the line between food and medicine will become increasingly blurred.

The market for functional foods is there, as witnessed by the growth in health-food products generally, but the perception of transgenic crops currently does not coincide with health foods, particularly outside the USA. The biotechnology industry is therefore looking to the developing world for a product (a 'killer application'), which through its modified properties will benefit millions who are in the greatest need. Such a product would provide a much-needed boost to the image of GM food.

Golden rice   Rice is the staple food of over three billion people. It is a good source of calories, but in its polished state lacks essential nutritional components such as vitamin A or ß-carotene, the chemical precursor of vitamin A. Each year, vitamin A deficiency is associated with more than one million childhood deaths, and it is the single most important cause of blindness among children. Transgenic rice with enhanced vitamin A may become the first high-profile functional food. It is called 'golden rice', because the polished grain (endosperm) is a pale yellow–orange colour.

Golden Rice™ was jointly developed by teams from the Swiss Federal Institute of Technology in Zurich and the University of Freiburg in Germany, led by Ingo Potrykus and Peter Beyer, respectively. Funding for Golden Rice development (US$100 million) was provided by the philanthropic Rockefeller Foundation, along with the Swiss Federal Institute of Technology, the European Community Biotech Program, and the Swiss Federal Office for Education and Science. A potential morass of intellectual property rights, involving over 70 patents, has not proved an obstacle, as the multinational companies involved, led by Syngenta (Novartis and AstraZeneca), have pulled together to get this flagship product to market (Kryder *et al.*, 2000).

Golden Rice was developed using 'first generation' transgenic crop technology, but the transformation was more complex. Instead of one gene construct (containing a transgene, promoter and marker gene), Golden Rice has three. Four metabolic reactions are needed to convert the chemical precursor called GGPP (geranylgeranyl diphosphate), naturally found in rice endosperm, into ß-carotene. Four different enzymes therefore need to be expressed by transgenes in order to make this four-step conversion. The first gene construct contains one gene from a daffodil

(*Narcissus pseudonarcissus*) and two genes from a bacterium (*Erwinia uredovora*), which express the first three enzymes; two antibiotic-resistance marker genes (hygromycin and kanamycin) from the bacterium; and promoters (including CaMV 35S). The second gene construct also contains transgenes for the first three enzymes, but without antibiotic-resistance markers. The third construct contains the gene expressing the fourth enzyme, again from the daffodil, along with an antibiotic-resistance marker gene (hygromycin) (Ye *et al.*, 2000).

Although heavily promoted as a 'gift' to developing countries, Golden Rice is likely to be aggressively marketed as a functional food in developed countries, particularly Japan. As a result of a deal in May 2000 between the rice's developers and Syngenta, the multinational will hold the commercial rights, but will license 'non-commercial' rights back to investors. Therefore, a two-tier system will operate, enabling vitamin A-enhanced rice to be sold at premium prices in developed countries, while being distributed free to governmental plant-breeding centres in the developing world. The Golden Rice deal is seen as the first in a series of 'Third World-friendly' moves by the biotechnology industry. If successful, Golden Rice may provide a model for other crops (Monsanto, 2000b). However, it remains unclear how well Golden Rice will perform in the field and how effective it will be in helping to cure vitamin A deficiency diseases (Chapter 12).

Medicinal products   In additional to functional foods, transgenic crops in the future will also be grown as a source of vaccines and therapeutic protein. Crops could be a highly cost-efficient way of producing edible vaccines against a range of animal or human pathogens. Vaccines have been produced on an experimental basis in thale cress (*Arabidopsis thaliana*), tobacco, banana (*Musa*), potato, cowpea (*Vigna*), rice and other plants. Hepatitis B vaccine, for example, has been produced in bananas. If this is given as a purée, a ten-hectare banana plantation could be enough to vaccinate all children under five in Mexico (Kiernan, 1996). Vaccines against diarrhoea, measles, yellow fever, diphtheria, cholera and polio are also being developed in transgenic plants. Small therapeutic proteins, such as the enzymes lacking in certain human disease, could be produced in transgenic crops. Alpha–1–antitrypsin has been expressed in transgenic rice grains, for example, which if successful in trials could provide treatments for liver and lung diseases (Kleiner, 1997b). Albumin, an important human blood product, has been produced in rubber trees. These will be high-value crops grown over relatively small areas. In the future, crops producing vaccines and pharmaceuticals may represent the most economically important application of transgenic crop technology.

The 'first generation' of transgenic crops was characterized by an arbitrary choice of transgenes. The main criterion was how much benefit the multinational companies could derive while making a product attractive to growers. To a certain extent the companies saw herbicide-resistant crops as ploughing the furrow in which a wider range of crops could be sown. However, this strategy backfired due to the large-scale opposition to transgenic crops and the food produced from them. The 'next generation' of transgenic crops promises to deliver more in the way of benefits. As we shall see in later chapters, however, a range of environmental and socio-economic factors may render genetic engineering not always the most appropriate technology for delivering the solutions it promises.

5 Invasion

Species become invasive when they spread and persist in new areas, with detrimental ecological consequences. Invasion biology is concerned with how and why this happens. In this chapter, several examples of 'worst case scenario' invasions are described, along with some principles derived from the literature of introduced species. In conjunction with field experiments, these can help predict the potential invasiveness of genetically modified organisms (GMOs).

Species introductions

An ecological invasion occurs when a species disperses and expands its range. In ecological theory, the removal of the normal regulating influences on a population is called ecological release. This has been proposed as a major factor in the success of invasive species, because the full range of regulating factors is unlikely to be present in the area being invaded. Population growth can therefore be rapid, due to the absence of natural enemies or the lack of competitors. Invading species typically occur at much higher population densities than they would in their home range (Elton, 1958).

Many of the world's worst pest problems have arisen from species introductions that have become invasions. These were the result either of accidental introductions or of deliberate releases that have had unanticipated ecological effects. Alien plants and animals introduced into the USA during the twentieth century have caused damage costing at least US$97 billion, according to a US Congress study. Worldwide, alien species present the second biggest threat to native species after habitat loss (Dobson and May, 1986; Kiernan, 1993; Todd, 2001).

Aquatic   Numerous examples can be cited of introductions that have had unforeseen and adverse environmental impacts. A classic case is the introduction of the Nile perch (*Lates niloticus*) into Lake Victoria in East Africa by the British in the 1950s, as a means of increasing food productivity. The carnivorous perch had little effect for about 20 years, but then its population density increased dramatically, until it accounted for around 80 per cent of the lake's biomass. As a result, the 400 species of indigenous fish, which themselves had once made up 80 per cent of the biomass, were reduced to around 200 species. Since the early 1990s, however, the numbers of Nile perch have started to decline, because they have eaten much of the available food and started to self-regulate their population by consuming their own offspring. This breakdown in the food chain is an indicator of ecological instability (Gophen *et al.*, 1995). A United Nations report recently estimated that 20 per cent of the world's freshwater fish species were threatened by invasive alien species (MacKenzie, 2001).

One of the most serious invasions of recent years resulted from the introduction of an alga called *Caulerpa taxifolia*, bred to look attractive in aquariums, into the Mediterranean Sea in the early 1990s, near Monaco's Oceanographic Museum. Wild-type *Caulerpa* species are native to Asia and are not invasive. Herbivores keep populations down in their native range. The laboratory-bred *Caulerpa* has a chemical that makes it inedible to herbivores, however, while its clonal mode of reproduction means that it can easily out-compete native Mediterranean species. Large areas that once supported a great diversity of plants and animals are now just monocultures of bright green *Caulerpa*. The alga now threatens to spread beyond the Mediterranean, either on ocean currents or through people emptying their fish tanks into the sea. A major operation was initiated in the USA, for example, to prevent a colony of *Caulerpa* causing ecological devastation along the Californian coast (Meinsz, 2001).

Many other aquatic invasions have resulted from unintentional human intervention. The Zebra mussel (*Dreissena polymorpha*) invaded the North American Great Lakes after being transported in the bilge water of cargo boats from the Black Sea. Populations have increased unchecked and caused damage in excess of US$5 billion, by choking the water intake and outlet pipes of power plants and other industrial facilities (Kiernan, 1993).

Plants   In terrestrial habitats, numerous plant introductions have had disastrous ecological effects, none more so than the kudzu vine (*Pueraria lobata*), which was introduced into the south-eastern USA from Japan in 1876. It was hoped that the plant would provide shade and contain soil erosion, but it now blankets the landscape. It has swamped everything with its rapid growth, and areas once rich in native plants have become monocultures of kudzu. Over seven million acres of Mississippi, Georgia and Alabama are under kudzu (equivalent to the size of Belgium). Freezing weather is the only natural deterrent to kudzu in its new range, and this prevents its march to the north.

In Britain, *Rhododendron* cultivars have escaped from gardens to invade wild habitats. It is not so much the rapid growth of rhododendrons that make them good invaders, but the mycorrhizal fungi associated with their roots, which release chemicals toxic to other plants. Therefore, native plants cannot recolonize ground cleared of rhododendrons for several years, whereas other rhododendrons can.

The characteristics most likely to turn plants into invaders are that they germinate in a wide range of conditions, are adaptable perennials, grow rapidly, flower early in the season, are self-fertilizing, are able to reproduce vegetatively (asexually), are good competitors, and have many seeds that disperse widely. No invader will possess all of these characteristics, but good invaders will have at least some of them. The lack of specialist herbivores, which control plant growth in the home range, is often an important factor in determining the severity of plant invasions. Human activity is the most important factor in the dispersal of plants (Tomiuk and Loeschcke, 1993).

Animals   Landscapes can be radically transformed by animal invasions. The European rabbit (*Oryctolagus cuniculus*), which originated in the Iberian Peninsula, wreaked ecological havoc after it was introduced into Australia in 1859 by Thomas Austin, on his extensive estate in Victoria. In its home range the rabbit rarely reaches the high population densities that were subsequently observed in Australia, because numbers are kept in check by predators, the dieback of food in winter, the harsh summers, and other factors. However, ecological release occurred in Australia, and the rabbit spread at a rapid rate. The reasons why the rabbit has been a good colonizer include its pre-adaptation to the climatic conditions in Australia, its arrival in the new area without its parasites and predators, the favourable disturbed environmental conditions created by woodland clearance and stock raising, and the depletion of native herbivores and predators (Thompson and King, 1993).

The population growth of rabbits in Australia was checked only by the introduction of the myxoma virus (myxomatosis) in the 1950s. Since then, European rabbit calicivirus disease (RCD) has further reduced rabbit populations, allowing native vegetation to recover. However, this virus escaped from an experiment conducted on an island off Australia in 1995, and has operated beyond the control of scientists from the Commonwealth Scientific and Industrial Research Organization (CSIRO), who released it. The CSIRO are now planning trials with a genetically engineered myxoma virus carrying an anti-fertility gene (Finkel, 1999).

The characteristics most likely to make vertebrates (fish, birds and animals) successful invaders include a large native home range in which they are abundant, a short generation time, a high degree of mobility, a gregarious life-style, being larger than related species, an ability to live in a wide range of physical environments, and an association with humans (Tomiuk and Loeschcke, 1993). In the case of birds and animals, herbivores are more likely to become invaders than carnivores (Crawley, 1987).

Biocontrol   Biological control involves the release of a species into a new area to control a pest insect, animal or plant (weed). Around 300 kinds of insects and pathogen have been introduced to control over 50 invasive plants, while around 1,000 predators, parasites and pathogens have been introduced against nearly 500 insect species. There have been many spectacular successes, and environmentally friendly biocontrol solutions to pest problems are increasingly sought. However, biological control remains a hit-or-miss affair, with only around half of introductions against weeds, and less than a third against insect pests, establishing and having any effect. There is a growing awareness that biocontrol is ecologically risky, with introduced species potentially becoming pests in their own right. Therefore, tougher standards of safety and more rigorous ecological risk assessments are now being applied to biocontrol agents (Malakoff, 1999).

Predicting invasions

Research into the potential of species to become invasive has until recently been a neglected area of study. The effects of invasion have traditionally been studied after the event. The emphasis now is on predicting invasions and working out management strategies to contain them so that native species can be protected. One area in which knowledge, gained by studying how and why invasions occur, can be applied is in the risk assessment and containment of GMOs in the environment.

Taking all the known instances of introductions, species become established in about one in ten cases, and of these one in ten becomes a pest. Therefore, there is around a 1 per cent chance of an introduced species gaining pest status. The main problem is that a number of these species were not previously considered as invasive. Major invasions can come out of the blue at any time, and are often hard to predict (Williamson, 1996).

Nevertheless, a number of approaches can be used to predict invasiveness. The most straightforward is to compile a list of species known to be weeds or pests in either their home range or elsewhere. In Europe, for example, oilseed rape (*Brassica napus*), sugar beet (*Beta vulgaris*), carrot (*Daucus carota*), oats (*Avena sativa*), lucerne (*Medicago sativa*), and a number of other crops are known to establish populations in the wild. Genetic modifications to these crops could result in increased invasiveness. A range of other crops, including onion (*Allium cepa*), maize (corn; *Zea mays*) and potato (*Solanum tuberosum*) do not form wild populations, and so the risk of invasive plants arising as a result of modifications to them is much lower (Bartsch *et al.*, 1993). A list of the characteristics of known invading species can also be compiled, to look for commonality as a predictive tool. For example, a particular reproductive mode (e.g. wind pollination) may be a key factor determining invasiveness within certain plant families. In addition, climate-matching can be done to assess the invasive potential of organisms in a particular region, based on their known climatic ranges. Meanwhile, mathematical models can be constructed using ecological factors to predict the course of an invasion. The above are all correlative methods.

A number of experimental approaches can also be used to predict invasiveness. These include the manipulation of environmental factors in growth chambers, greenhouses or field experiments, and assessments of how individuals of a species survive when deliberately released beyond their native range. Experiments on the potential invasiveness of genetically modified plants sown in uncultivated areas, compared to unmodified plants, have sparked new interest in invasion biology generally (Mack, 1996).

Assessing the invasiveness of transgenic plants

One of the concerns relating to transgenic crops is whether they could become more invasive than non-transgenic cultivars, and thereby gain the potential to become weeds in other crops or in uncultivated habitats. In crop plants, for example, modifications could increase invasiveness if they altered the number or viability of seeds produced. However, field studies with transgenic plants that measure such characteristics are still relatively rare, particularly those that include more than one generation.

Invasiveness is related to reproductive fitness, which can be quantified as the number of offspring produced by an individual. Increased reproductive fitness can lead to increased invasiveness, although this need not necessarily be the case, owing to a range of other ecological factors. Invasion is a complicated process, as reproductive fitness is affected by changing ecological conditions over time and within different habitats. For species to become invasive, populations need to disperse and establish outside of their previous range. They have to find and occupy space, and out-compete any indigenous organisms that use the same resources. Invasiveness is dependent on dispersal and colonization rates, the number of individuals involved, whether arrival is a repeated or a one-off event, the mode of reproduction, and the degree of human interference. Chance events also play their part. In addition, certain plant communities are more susceptible to invasion than others.

The inadvertent spread of crop seed is an important route by which transgenic plants could invade new areas (Crawley, 1996). Seed can be spread by birds, on the fur of animals, or by human activity. Seed travels long distances from seed merchants to farmers, giving plenty of opportunity for spillage in transit. The seed trade has been linked with inadvertent seed dispersal in the USA (Mack, 1991). In Britain, wild oilseed rape has established along roadside verges via seed spillage. Its distribution around London's orbital motorway (M25) has been correlated with the transit routes of seed-carrying lorries (Crawley and Brown, 1995). If populations of transgenic crop plants become established outside cultivated areas in this way, they could become an important source of pollen, capable of creating hybrid plants containing transgenes (Chapter 6). Weed sugar beet is a serious problem in cultivated sugar beet, which is a biennial crop. In a French study, the pollination of cultivated sugar beet by nearby weed beet resulted in hybrid weed sugar beet. The disturbed habitats around fields have selected for weed sugar beet that flowers after only one year instead of two, making it potentially more invasive (Boudry *et al.*, 1993).

A research group called PROSAMO (Planned Release Of Selected And Modified Organisms), headed by Mick Crawley and funded by Britain's Department of Trade and a consortium of biotechnology companies, was established at Imperial College, London, in the early 1990s to study the invasiveness of transgenic crops. In the main part of the study, oilseed rape, a crop known to colonize non-agricultural land, was grown over three seasons, in three climatically distinct regions, in four habitats. Unmodified plants were compared with transgenic plants that had either an antibiotic-resistance marker gene (kanamycin) or a full herbicide-resistance (glufosinate) gene construct (also containing the marker gene). Plants were monitored in cultivated and uncultivated habitats, the latter mimicking the situation for oilseed rape plants when they escape into undisturbed natural habitats. Vertebrate herbivores (mainly rabbits) and plant competitors were eliminated, using fencing and hand hoeing, while pesticides were used to control fungal and invertebrate pests. No herbicides were sprayed on the plants.

The PROSAMO researchers took measurements each year for the survival of individuals, the proportion of seeds germinated, the proportion of seedlings that died over winter, and the mean number of seeds produced per seed that germinated. These measurements were fed into an equation to give rates of population increase. Substantial variation was reported in seed survival, plant growth, and seed production between sites and across experimental treatments; but there was no indication of change in invasive potential. Transgenic oilseed rape appeared to be no more persistent in uncultivated habitats than its conventional counterpart. Differences in rates of population increase in uncultivated habitats were mainly influenced by interspecific competition: competition amongst oilseed rape plants for resources (Crawley *et al.*, 1993).

In a ten-year study, completed in 2001, Mick Crawley's team grew four different crops (oilseed rape, potato, maize and sugar beet) in uncultivated habitats and found no evidence that transgenic plants were more invasive than their conventional counterparts. The sites were monitored each year, and transgenic varieties persisted for no longer than their conventional counterparts, confirming the earlier experiments. Population sizes of all crop types declined after the first year, due to competition from native perennial plants. The results for the four crops suggested that they would not normally survive for long outside cultivation (Crawley *et al.*, 2001).

It was the modification process itself that was being tested in the PROSAMO experiments, because no antibiotics or herbicides were applied to the plants. In some cases, transgenic plants did less well than non-transgenic plants; but overall, no substantial costs to the plants appeared to be incurred by the expression of antibiotic-resistance or herbicide-resistance transgenes in situations where they provided no selective advantage. Under selection pressure from herbicides, however, transgenic varieties would probably persist longer than non-transgenic varieties, although herbicide spraying is uncommon on weedy oilseed rape and wild populations of other crops in uncultivated situations.

Reproductive fitness advantages have been shown for transgenic over non-transgenic crops, but only under specific conditions. In experiments with transgenic insect-resistant *Bt* oilseed rape, the ability to tolerate herbivore attack conferred fitness advantages compared to non-transgenic plants, but only when medium to high levels of defoliation occurred (Stewart *et al.*, 1997).

Transgenic sugar beet, modified to be resistant to the viral pathogen Beet Necrotic Yellow Vein Virus (BNYVV), similarly had higher seed productivity than an unmodified sugar beet from the same breeding stock under virus-infestation conditions, but not under virus-free conditions. Sugar beet is normally not competitive enough to become an invasive weed, and it was concluded that the risks associated with transgenic and conventional virus-resistant sugar beet were equivalent (Bartsch *et al.*, 1996). Transgenes therefore confer fitness advantages only in the presence of relatively high selection pressure; for example, due to the pests and diseases that GMOs have been modified to resist. These selection pressures may only occasionally be present in uncultivated habitats, but if they are, increased invasiveness can result.

In agricultural ecosystems, weeds derived from transgenic herbicide-resistant crops are a particular concern. A project to obtain resistance to the herbicide sulfonylurea in oilseed rape, for example, was abandoned when it was realized that the oilseed rape could become a weed in wheat, where it would be resistant to the most important herbicide used in that crop (Oxtoby and Hughes, 1990). As a general precaution, plants modified for herbicide-resistance should remain susceptible to at least one major group of herbicides. The reproductive fitness advantages of a herbicide-resistance transgene will become evident only when that particular herbicide is used as a weed management tool. In the following chapter, it will be seen how ecological risk is increased as a result of herbicide-resistance transgene spread in pollen.

The 'first generation' of transgenic crops has population dynamics that do not differ markedly from equivalent non-transgenic crops (Crawley, 1999). However, increased invasiveness may be more of a concern in the future, with tolerance of a wider range of pests and diseases, and especially with transgenic crops modified for drought or saline-tolerance, or an enhanced ability to fix nitrogen. Such plants will have the ability to extend their ranges. Although the risks are likely to be relatively minor in comparison to the impacts of introduced species generally, releases of transgenic crops or other GMOs in the future could lead to the creation of ecologically damaging invasions.

6 Genetic pollution

The genes from a transgenic plant can spread into the environment via a number of routes. Pollen that is airborne or carried by insects can potentially fertilize sexually compatible wild relatives or other cultivars to produce hybrid plants containing transgenes. Seed can be spilt in transit, carried by animals, or blown on the wind. In addition, genes may pass from plants into the soil, where bacteria, fungi and other microorganisms can potentially pick them up. In this chapter, we will see that gene escapes from many transgenic crops are inevitable once they are grown on a commercial scale. The term 'genetic pollution' has come into common usage to describe the unplanned spread of genetically modified organisms (GMOs) and the transfer of transgenes to unmodified organisms.

Gene flow from transgenic crops

During sex in flowering plants (Angiosperms), pollen grains released from structures called anthers, at the tip of the stamen (male part of the flower), become trapped on the stigma (female part of the flower). A pollen tube grows down through the stigma to the ovule, where fertilization occurs. A viable seed is the product of fertilization. This can occur between flowers on the same plant (self-fertilization) or different plants (cross-fertilization). Flowering plants are designed mainly to be cross-fertilized. As noted earlier, although species are defined in terms of interbreeding individuals, the boundaries between species are often ill-defined. In flowering plants, interspecific hybrids can be common.

Scientists who study gene flow in the environment talk in terms of introgression: the spread of genes from one species into the gene pool of another by interspecific hybridization. In this sense 'genetic pollution' is nothing new, as domesticated plants have always exchanged genetic material with related species. Weed races have evolved from most of the major cultivated crops, and these weeds can still cross with those crops.

Gene flow is the movement of genetic information among individuals, populations or taxonomic groups. Potential gene flow is the movement of seed or pollen away from a source. This dispersal of genes, however, amounts to nothing if seeds fall on stony ground or no related plants are available to receive the pollen. Actual gene flow is a measure of the number of individuals produced (from seed) and amount of fertilization (from pollen) as a function of distance from source. Gene flow can be measured directly or indirectly. Direct observations of seed or pollen dispersal in the field give an estimate of potential gene flow. Indirect observations using genetic markers give an estimate of actual gene flow (Raybould and Clarke, 1999).

The inadvertent spread of seed in transit is an important route of gene flow, as noted previously. Better containment of seeds during transport could cut down on this source of 'genetic pollution'. Seeds can also be transported on the wind and by animals and birds, particularly if careless handling occurs during sowing or harvesting. Improved crop hygiene could reduce stray seed, but it would be very difficult to account for it all. This seed can grow into weeds in later crops on the same land (volunteer plants) or establish populations outside cultivated areas.

Gene flow in pollen   Pollen can be freely dispersed from transgenic plants, and this pollen can potentially travel for considerable distances. In small-scale field trials, the amounts of pollen involved were relatively small. The real extent of transgene spread via pollen has become apparent only since transgenic crops have been commercially cultivated.

Two main areas of concern have arisen as a result of gene flow in pollen from transgenic crops. The first is that pollen will fertilize wild relatives to create weedy hybrids with undesirable characteristics. The second is that transgenic crops will fertilize non-transgenic cultivars and reduce their 'genetic purity'. This is a particular problem in crops grown for their seed or for their organic and non-transgenic status.

Gene flow has been studied for a number of years in the context of seed crops, grown for the seed they produce. Minimum isolation distances need to be established to maintain acceptable levels of varietal purity in seed crops. For example, brassicas, which are mainly outbreeding (matings between distantly related individuals are favoured within a species) and insect-pollinated, have a minimum isolation distance of 900 metres. Cereals such as wheat and oats, which are self-fertilized and wind-pollinated, and sugar beet, which is cross-fertilized and insect- and wind-pollinated, have minimum isolation distances of 180 metres and 3,200 metres, respectively (Levin and Kerster, 1974; Levin, 1984). However, field trials conducted to establish minimum isolated distances have been carried out in a limited range of weather conditions, and long-distance dispersal has not been looked for. Such estimates are now considered on the low side as indicators of pollen movement, in the light of more recent research, and may give a misleading picture if used for the risk assessment of transgenic crops. The recommended isolation distance for radish of 300 metres, for example, could allow anything up to ten per cent of 'foreign' cross-pollination (Ellstrand *et al.*, 1989). Meanwhile, pollen grain clusters can occur in some wind-pollinated species, concentrating gene flow in distant patches. Furthermore, small amounts of pollen that travel long distances might have a disproportionate importance in terms of gene flow and hybridization. Many factors influence gene flow and its analysis is far from straightforward (Tonsor, 1985; Lanner, 1966; Raybould and Gray, 1993).

Nevertheless, generalized estimates of the percentage of pollen remaining airborne with distance downwind from transgenic crops, in light to moderate winds, can be obtained from the published literature. In one review, the percentage of maize (corn) pollen remaining airborne, compared to concentrations 1 metre from the crop, was about 2 per cent at 60 metres, 1.1 per cent at 200 metres, and 0.75–0.5 per cent at 500 metres (Emberlin *et al.*, 1999). Estimates of this type can provide a rough guide, but many factors can affect pollen movement. Pollen can be transported over much longer distances than those usually considered for crops, under certain weather conditions. In storm fronts, for example, particles in the air can be lifted up and along for many kilometres. This long-range dispersal of pollen will be discussed later in the context of tree pollen (Chapter 9).

The potential for pollen to cause gene flow through hybridization depends on a number of factors, including the amount of pollen produced, its longevity, modes of dispersal (wind, insects), the distance between donor and recipient plants, the relative abundance of plant populations, the height of plants, the characteristics of the intervening vegetation (topography), the prevailing climate and weather (wind speed, temperature), and the degree of compatibility between the plant genomes (Daniell, 1999). The amount of transgene flow via pollen spread varies with the type of crop grown, and can differ for the same crop depending on the prevailing environmental conditions. There is no reason to suppose that transgenic pollen behaves in a fundamentally different way from non-transgenic pollen (Raybould and Gray, 1993).

Most pollen loses its viability relatively quickly, especially under unfavourable environmental conditions, such as high temperatures, high humidity or strong light. Most maize pollen loses its viability within twenty minutes, particularly in hot weather, although some remains viable for an hour or two, or even a day under moderate conditions. Exceptionally, in very cool weather, maize pollen has remained viable for up to nine days (Jones and Newell, 1948; Purseglove, 1972; Emberlin *et al.*, 1999). Cereal pollen generally has a short viability time in the field. The pollen from cultivated rice mostly loses its viability within three to five minutes, although the pollen from wild rice is viable for a little longer (Koga *et al.*, 1971). Pollen longevity in wheat and rye has been reported as 45 and 220 minutes, respectively (Fritz and Lukaszewski, 1989). There is little published data on the viability of oilseed rape pollen; some reports suggest it can remain viable for up to three days, although most will remain viable for much shorter periods. If pollen loses its viability, how far it travels is irrelevant in terms of gene flow, although even short viability times can lead to fertilization at some distance from the source plant.

The sticky pollens of predominantly insect-pollinated crops, such as cotton, generally have greater longevity than wind-pollinated crops, but this rarely exceeds one day (Richards, 1986). The viability of pollen carried by insects declines as the number of flowers visited increases. For honeybees (*Apis*), the majority of oilseed rape pollen that is picked up will be deposited on one of the next fifteen to twenty flowers visited. For bumblebees (*Bombus*), most of the cross-pollination (around 91 per cent) occurs on the next four flowers visited. Therefore, although these pollinating insects commonly range over distances of two kilometres or more, most cross-pollination is likely to be relatively localized (Cresswell, 1994). There is evidence, however, that honeybees can carry transgenic oilseed rape pollen over distances of at least four kilometres (Chapter 7).

Crop-weed hybridization

Robust and sophisticated DNA typing ('fingerprinting') methods are now available that can identify the actual gene flow from one plant to another with more sensitivity and accuracy than ever before. This has radically changed the way that genetic purity is perceived. Whereas a certain level of hybridization in seed crops would previously have been undetectable, and therefore was accepted, new testing methods can show up every case of cross-fertilization. To confirm gene flow, it is necessary to observe the transfer of genetic material, in the form of a genetic marker, from one individual or group to another. Therefore, both the potential donor and recipient plants must be genetically typed so that distinguishing features in their DNA can be quantified. If a particular DNA sequence is found in a recipient plant but not in its mother, then gene flow has occurred, as this sequence has been derived from incoming pollen.

If the number of potential donor plants is small, then DNA can be traced back to the plants from which the pollen came. In principle, this sounds reasonably straightforward, but in commercial situations with gene flow into a non-transgenic crop, transgenes will be competing with a much larger concentration of pollen from within the conventional crop itself. This dilution effect means that the proportion of seeds acquiring a transgene may be very small. A number of highly sensitive DNA analysis techniques are now available. A technique called ISSR analysis, for example, is used to type the DNA of grass species and oilseed rape (Squire *et al.*, 1999).

Hybridization and crop type   A number of transgenic crops grown commercially are unlikely to be a problem in terms of crop-weed hybridization. Soyabeans (*Glycine max*) and cotton (*Gossypium*), for example, which account for around 90 per cent of all transgenic crops grown in the USA, are non-native and have no wild relatives in the areas where they are cultivated in that country. Similarly, maize (*Zea mays*), barley (*Hordeum vulgare*), potato (*Solanum tuberosum*) and tomato (*Solanum esculentum*) present minimal risks in terms of crop-weed hybridization in the USA and Europe. Transgenes from these crops have nowhere to go outside agricultural systems, although cross-pollination with other crop cultivars may be a cause of concern. On the other hand, there are many transgenic crops, including oilseed rape (*Brassica napus*) and other brassicas, sugar beet (*Beta vulgaris*), sunflowers (*Helianthus annuus*), rice (*Oryza sativa*), oats (*Avena sativa*), ryegrass (*Lolium*) and other cereals and pasture grasses, that are grown in areas where many wild relatives occur with which they can potentially exchange genes (Brookes, 1998; Keeler *et al.*, 1996). In the UK, Raybould and Gray (1993) assessed the risks that a range of plants posed with respect to forming wild-type or weedy hybrids. The plants identified as being at high risk included sugar beet, cabbage (*Brassica oleracea*), ryegrass, white clover (*Trifolium repens*), lucerne (*Medicago sativa*), carrot (*Daucus carota*), plums (*Prunus domestica*), apple (*Malus pumila*), Scots pine (*Pinus sylvestris*) and poplar (*Populus*).

The quantification of gene flow from transgenic crops to wild-type relatives, which are often common weed species, has now been demonstrated in a number of experimental field studies. Many of these have been done with brassicas, which readily hybridize with wild relatives, can persist as free-living weeds, and have pollen that is spread by both wind and insects (Snow and Morán Palma, 1997). Many wild relatives of brassica crops commonly occur around agricultural land.

Oilseed rape   Oilseed rape (*Brassica napus*) originated as a hybrid of *Brassica oleracea* and *Brassica rapa* (*B. campestris*). Although both its parent species are typical of the brassicas in being outbreeding, oilseed rape is predominantly self-pollinated, with additional cross-pollination mediated by both wind and insects. Nevertheless, it is known to hybridize with at least fourteen weed species, including wild turnip (*Brassica rapa*) and wild radish (*Raphanus raphanistrum*) (Daniell, 1999; Keeler *et al.*, 1996; Scheffler and Dale, 1994).

Wild turnip is a widespread weed that thrives in disturbed soils around agricultural land and is commonly found close to oilseed rape fields. In one of the first transgenic crop–weed studies, conducted in Denmark, spontaneous crop–weed hybridization between transgenic oilseed rape, modified to be resistant to the herbicide glufosinate (Basta®), and wild turnip was demonstrated (Jørgensen and Andersen, 1994). Weed-like and herbicide-resistant hybrid plants were produced, which resembled wild turnip. The presence of the transgene in these hybrids was confirmed by genetic analysis. Herbicide-resistant hybrid plants were found at the site of the experiment during the following spring among plants that had germinated from seeds shed at the time of the previous year's harvest. Field experiments also demonstrated spontaneous backcrossing between the oilseed rape–wild turnip hybrids and wild turnip. The hybrid plants had high pollen fertility, produced abundant viable seed, and had a higher reproductive fitness (seed produced per individual) than parent wild turnip plants. Spontaneous hybridization has also been demonstrated between oilseed rape and the wild relatives *Brassica juncea*, *Raphanus raphanistrum* and *Hirschfeldia incana*. Rikke Bagger Jørgensen and his colleagues, at the Risø National Laboratory in Roskilde, Denmark, concluded from their results that a rapid spread of genes from oilseed rape to weedy relatives appeared to be inevitable under normal agricultural conditions (Mikkelsen *et al.*, 1996; Jørgensen *et al.*, 1996; Jørgensen *et al.*, 1998; Jørgensen, 1999).

Field trials with transgenic oilseed rape surrounded by wild radish (*Raphanus raphanistrum*) were conducted in France to study spontaneous gene flow. A transgene, which confers resistance to the herbicide Basta (glufosinate), passed from oilseed rape to radish and persisted for four generations in oilseed rape–wild radish hybrids. In the fourth generation, 20 per cent of the hybrids retained the gene for herbicide-resistance, although the authors of this study concluded that under normal agricultural conditions such gene flow would be relatively rare (Chèvre *et al.*, 1997).

Pollen from a non-transgenic oilseed rape cultivar was shown to travel long distances to fertilize wild (feral) oilseed rape, in a study conducted by the Scottish Crop Research Institute (SCRI). Airborne pollen densities were monitored at various distances from isolated oilseed rape fields (3–10 hectares in size) over a three-year period. Pollen densities at 360 metres from the crop were around 10 per cent of densities recorded at the field margin. Pollen counts of 0–22 pollen grains per m³ were observed 1.5 kilometres from source fields, while pollen from the isolated fields consistently reached wild populations of oilseed rape up to 2.5 kilometres downwind (Timmons *et al.*, 1995; Timmons *et al.*, 1996). Therefore, significant quantities of viable oilseed rape pollen travel over long distances.

The SCRI team proceeded from studying gene flow in non-transgenic oilseed rape to larger-scale studies with transgenic plants containing marker genes. In these experiments, small plots of male-sterile oilseed rape were used as recipient plants, acting as biological pollen detectors. Male-sterile bait plants do not produce their own pollen and represent a 'worst-case scenario' with respect to gene flow, because they cannot be self-pollinated and are 100 per cent fertilized by incoming pollen. These studies confirmed that significant amounts of transgenic pollen could be transported over thousands of metres.

Oilseed rape pollen has a greater capacity for long-range dispersal than suggested by small-scale field trials. The difficulty of extrapolating between different scales makes the study of gene flow difficult. Ecologists think in terms of the connectivity of interdependent patches or fragments of plants within a landscape. How these patches scale with respect to the landscape is far from straightforward. Large pollen sources, such as crop fields, interact on a regional scale to determine gene flow and may need to be studied at this level (Squire *et al.*, 1999). In another SCRI study, in an area of conventional commercially grown oilseed rape west of Dundee, genetic contamination of seed from a transgenic crop growing some four kilometres away was observed. Airborne pollen was still viable after travelling long distances, and was shown to pollinate flowers. The SCRI studies showed that farm-to-farm transgene spread in oilseed rape will be widespread under the cultivation conditions found in Europe, while escaped transgenes could persist in wild relatives. Honeybees were found to be much less important than wind in transferring pollen between cultivated oilseed rape plants, but were shown to be important carriers of pollen between crops and wild relatives (Ramsay *et al.*, 1999; Thompson *et al.*, 1999).

In a large-scale study of cultivated oilseed rape and its weedy relatives along a stretch of the River Thames in south-east England in 1998, one hybrid was observed from 505 wild turnip plants screened (Wilkinson *et al.*, 2000). It is difficult to extrapolate from this instance of cross-pollination, but the result backed up the growing realization that gene transfer is inevitable (Sample, 2000). The striking difference in hybridization rates reported in ecological studies with oilseed rape and its wild relatives, between 10 and 90 per cent, and the variability in hybridization observed makes quantifying risk difficult. However, weeds growing within oilseed rape fields are more likely to hybridize with the crop than individuals in populations adjacent to fields, while risk will generally decline with distance from a transgenic crop (Gliddon, 1999). The ecological risks will depend on the transgenes involved. To avoid undesirable gene flow, the movement of seed and pollen will have to be managed far more carefully in the future, while more co-ordination of oilseed rape planting between farms will be required (MAFF, 1999).

Vegetables and other crops In experiments with small plots of wild radish (*Raphanus sativus*), an average of ten per cent of seeds produced were as a result of fertilization from outside the plots (Devlin and Ellstrand, 1990; Ellstrand *et al.*, 1989). Gene transfer from cultivated radish (*R. sativus*) to wild weedy relatives has been detected over distances of up to one kilometre from field plots (Klinger *et al.*, 1992). The offspring of these crosses showed 'hybrid vigour', and produced more seed than normal weeds. This suggests that populations of these hybrids may establish rapidly in the environment (Klinger and Ellstrand, 1994).

Studies on the gene flow between a range of other crops and their wild relatives have also shown the potential for weedy hybrids containing transgenes to become established. Lettuce (*Lactuca sativa*) has around nineteen wild relatives including economically important weeds in many regions of the world that it readily hybridizes with, for example *L. serriola*, *L. virosa* and *L. saligna* (Keeler *et al.*, 1996).

A low frequency of pollen dispersal from a field trial of transgenic potatoes (*Solanum tuberosum*) was found in one study, with surrounding conventional potatoes. No transgenic hybrid seed was recovered from the unmodified plants further than 4.5 metres from the transgenic plants (Tynan *et al.*, 1990). However, a higher frequency of gene flow was demonstrated in another study of cross-pollination between transgenic and unmodified potatoes. A genetic marker was found in 72 per cent of unmodified potato plants in the immediate vicinity of the transgenic plants, and in around 35 per cent of unmodified plants grown up to 1,100 metres away (Skogsmyr, 1994).

The gene flow from transgenic strawberries (*Fragaria*) and sunflowers (*Helianthus annuus*) to wild relatives was high in field experiments that used marker genes present in pollen. More than 50 per cent of wild strawberries growing within 50 metres of a strawberry field were found to contain marker genes from the cultivated strawberries (King, 1996). In field experiments in Canada, high frequencies of crop-specific marker genes were found in wild sunflowers planted around cultivated sunflower fields.

Up to 42 per cent of the progeny from wild sunflowers were found to be hybrids containing crop marker genes. These transgenes persisted for at least five generations in wild sunflower populations (Linder *et al.*, 1998). Hybridization and gene flow (introgression) has also been recorded between cultivated sunflower and its wild relative *Helianthus petiolaris*, although at a lower frequency than for wild sunflowers (Rieseberg *et al.*, 1999).

Most of the research to date on gene flow from transgenic crops has been conducted in developed countries. However, the ecological risks due to crop–weed hybridization are theoretically greater in many parts of the developing world, particularly in the tropics, where biodiversity is greatest. The home ranges of few major crops occur in the USA, so there will be relatively few wild relatives to accept pollen from transgenic crops. Sugar beet and most brassica crops originated in the Mediterranean, and so transgenic sugar beet and oilseed rape can hybridize with wild relatives in Europe. The incidence of hybridization between many other transgenic crops and wild relatives can be expected to be higher in South and Central America, Africa and Asia. These areas are the focus of much crop diversity and have most of the wild ancestors and relatives of those crops. The centres of origin of maize, potato and tomato are in Central and South America; many cereal crops originated in East Africa; rice originated in South-East Asia; soyabeans originated in China. By October 2001, the finding of illegal, and therefore unregulated, plantings of transgenic maize in Mexico and transgenic cotton in India raised concerns that transgenes might soon be out of control in biodiverse areas of the developing world (Jayaraman, 2001).

Rice   Asia is the origin of rice (*Oryza sativa*) and is also home to numerous related species. Rice is basically self-pollinating, but natural outcrossing has been reported at rates of up to 5 per cent. Outcrossing levels are generally higher in Indica rice cultivars and wild species than in Japonica rice cultivars (Oka, 1988). There are 22 species of wild rice and these are known to form hybrids with cultivated rice (Oka and Chang, 1961). Of the many related wild species, *Oryza rufipogon* (the wild progenitor of *O. sativa*) and *O. nivara* are particularly abundant in many parts of Asia, and readily hybridize with cultivated rice. Wild rice, with which cultivated rice can potentially hybridize, also occurs in Africa (*Oryza barthii* and *O. longistaminata*), America (*O. glumaepatula*) and Oceania (*O. meriodinalis*). Given the extensive areas under rice and the persistence of hybrids in the field, gene flow from transgenic rice to wild relatives appears almost certain. Pollen from rice is only generally viable for minutes, but this is sufficient to fertilize wild rice in the vicinity of farmland. Wild rice is a superior competitor, due to its vigorous growth and height, and has become a serious weed problem in many areas, particularly where farmers have switched from the transplanting of seedlings to the direct sowing of rice seed.

Cultivated rice is being genetically engineered for a variety of traits. Rice is being made insect-resistant using a *Bt* toxin gene (*Cry1A*) against stem borers, and herbicide-resistant. Modifications that lie further in the future include drought-resistance and tolerance of high salinity. Transgenic cultivars could therefore make a valuable contribution in raising rice yields. However, if transgenes became established in wild rice or related species they might result in hybrids that are more competitive or invasive, which could ultimately be detrimental to rice and possibly other crops. More research is required on what determines the distribution and abundance of *Oryza* species; their genetic diversity; the extent of gene flow among wild, weedy and cultivated rice; and the relative fitness of hybrids of transgenic rice and wild and weedy relatives (Cohen *et al.*, 1999). In the meantime, the International Rice Research Institute (IRRI) in the Philippines has conducted field trials of transgenic rice resistant to bacterial blight, and China is pushing ahead with commercial releases of transgenic rice.

Gene flow in the soil   Genetic material can be transferred by routes other than seed or pollen spread, for example by leaching from the root system. Transfer of DNA from plants to viruses has already been mentioned (Chapter 3), while gene transfer can also occur from plants to bacteria and fungi. In one experimental study, transgenic oilseed rape, black mustard (*Brassica nigra*), thorn-apple (*Datura stramonium*) and sweetpea (*Lathyrus odoratus*), all containing introduced antibiotic-resistant genes, were grown together with the fungi *Aspergillus niger*. In each case, the fungi incorporated the antibiotic-resistant gene (Hoffmann *et al.*, 1994). Commercially grown transgenic crops have routinely contained antibiotic-marker genes, which could be incorporated into soil microorganisms via such mechanisms.

Gene flow and hazard   Given the accumulating knowledge on transgene flow, it should be assumed that if GMOs are released into field situations, transgenes will escape into the wider environment by one or several routes. The question then is not whether transgenes will spread to the wider environment, but to what extent it matters. If plants grown from hybrid seed do not persist in the environment, or the transgenes involved do not present an ecological or agricultural hazard, then they may be of little concern. If the hybrid plant does present a hazard, then no amount of barrier zone will ultimately protect the environment, and the transgenic crop in question should probably not have been released into the environment in the first place.

Persistence of crop–weed hybrids It should be borne in mind that crop–weed hybridization is a ubiquitous process. All crops can potentially transfer genes to compatible wild relatives. Numerous plant species commonly form hybrids, exchanging genes in the process. The term 'genetic pollution' has often been used as if there is an inherent value in the purity of gene pools, which are sullied by the introgression of transgenes, but this is an emotive connotation that sits awkwardly with the reality of natural processes. Hybrids are created naturally. Hybrid plants are often sterile, and relatively few establish populations that persist in the environment to become weeds in agricultural ecosystems. But the cases in which fertile offspring do produce stable populations of interspecific hybrid plants, which persist for many generations, may be of great significance. Crops such as oilseed rape and oats are themselves the descendents of hybrids. In the British flora, comprising around 1500 native and 1000 alien species, about 770 hybrids have been described (Raybould and Gray, 1993).

In a worst-case scenario, a hybrid may become highly invasive, becoming a weed in crop situations or extending its range and putting rare and vulnerable native plant species at risk. Johnsongrass (*Sorghum halepense*) is a noxious weed of this type, formed as a hybrid of cultivated sorghum (*S. bicolor*) and a wild relative (*S. propinquum*) (Snow and Morán Palma, 1997). Weed beet, a hybrid of sugar beet and wild beet, has become a serious weed problem in sugar beet in Europe. As noted previously, hybrid weed beets have evolved early flowering characteristics, becoming annual rather than biennial. Within any sugar beet field there will be a small proportion of bolters – plants that flower a year early and release pollen. Gene flow from bolters to weed beet has been demonstrated in France. Transgenes, for example for herbicide-resistance, in crop–weed hybrids could increase weed problems in sugar beet crops (Vigouroux *et al.*, 1999).

Even though hybridization occurs, it does not mean that hybrids will become successfully established in the environment. To become invasive, crop–weed hybrids containing transgenes need to gain a fitness advantage from their novel genetic make-up. It has been widely assumed that crop–weed hybrids would have lower reproductive fitness than wild-type weeds, thus minimizing the danger of transgenes establishing in the environment. However, this is not always the case.

The reproductive fitness of first generation (F_1) hybrids of oilseed rape and wild turnip was significantly higher than for wild turnip. Reproductive fitness was quantified as offspring produced per individual, by scoring seed development, seed survival in the field, and subsequent pod and seed set. The second generation (F_2) hybrids had a reduced fitness relative to both parents, although variability was high, with some individuals being as fit as their parents (Hauser *et al.*, 1998a; Hauser *et al.*, 1998b). In another experiment, hybrids (F_1) of oilseed rape and wild turnip were also shown to be fitter than the wild turnip parent. These hybrids readily backcrossed with weedy parents under field conditions (Jørgensen *et al.*, 1998). 'Hybrid vigour' has also been noted for hybrids of cultivated radish and a weedy relative (Klinger and Ellstrand, 1994). Therefore, hybrids derived from transgenic crops and weedy relatives could become established, particularly in fallow fields or transient habitats around agricultural land that favour weed populations.

Herbicide-resistant weeds   The main concern that has arisen in agricultural situations, given that most transgenic crops are herbicide-resistant, is that gene flow will create herbicide-resistant weeds. All the cases of crop–weed hybridization considered in this chapter have the potential for creating weeds that may become more difficult to control in field situations if herbicide-resistance transgenes are involved. Goatgrass is an important weed problem in winter wheat in the western USA, for example, and the use of transgenic herbicide-resistant wheat cultivars would enable more effective weed control, using herbicides for grass weeds that might normally harm wheat. However, wheat (*Triticum aestivum*) can hybridize with jointed goatgrass (*Aegilops cylindrica*) in the field, although the hybrids have relatively low seed viability. Therefore, there is the potential for herbicide-resistance transgenes to be transferred into hybrid weeds to the detriment of weed control in wheat. The strategy could be viable, but would involve strict management (Seefeldt *et al.*, 1999). In all cases, the benefits of herbicide-resistance for weed controls must be balanced against the potential costs of creating herbicide-resistant weeds.

Genes from transgenic oilseed rape can enter populations of wild oilseed rape and weedy relatives. Populations of weedy oilseed rape are known to persist for over ten years. But to what extent will these weeds take advantage of potential niche modifications due to acquired transgenes? Transgenes need to confer a selective advantage if they are to aid in the persistence and spread of a population. The acquisition of herbicide-resistance, for example, would give hybrids a competitive advantage only if that particular herbicide were sprayed on them. Even then, other herbicide groups or weed control practices would still be available to control them (Squire *et al.*, 1999). A study of wild oilseed rape by the SCRI in 1995 found that 44 per cent of populations had been subjected to control measures, although only 14 per cent of these controlled populations were treated with herbicides, compared to 89 per cent being mowed. The selection pressure favouring the rapid spread of herbicide-resistance transgenes is therefore present in relatively few wild oilseed rape populations (Timmons *et al.*, 1996).

Ecological theory suggests that herbicide-resistance is costly to maintain in the absence of herbicides. Without the selection pressure of a herbicide, a herbicide-resistance gene leads the plant to divert valuable resources into making an unnecessary protein, which may reduce its reproductive fitness. There is data to support this hypothesis; for example, a cultivar of oilseed rape bred by conventional methods to be resistant to triazine herbicides was found to have a decreased seed-yielding ability (Beversdorf *et al.*, 1990). The development of transgenic plants has enabled this hypothesis to be more thoroughly tested (Bergelson and Purrington, 1996).

Joy Bergelson and her colleagues used transgenic thale cress (*Arabidopsis thaliana*), engineered with a gene that conferred resistance to the herbicide chlorsulphuron, to test whether any fitness costs were associated with herbicide-resistance. They found that transgenic thale cress had a 34 per cent reduction in life-time seed production compared to unmodified thale cress. In this case, an introduced herbicide-resistance gene caused a fitness cost (Bergelson *et al.* 1996; Bergelson, 1994). A number of studies with transgenic crop–weed hybrids have, however, shown that they produce the same number of seeds as wild-type weeds (Anon., 1998; Klinger and Ellstrand, 1994). For example, no obvious fitness costs, in terms of survival or seed production, were observed in wild turnip that had acquired glufosinate-resistance transgenes from oilseed rape, even in the absence of herbicide treatments (Snow and Jørgensen, 1999). Overall, the results with transgenic crop–weed hybrids suggest that herbicide-resistance transgenes might not be as costly to plants as predicted by ecological theory (Hails, 2000).

If transgenic plants were to donate more pollen to wild-type relatives than unmodified plants, it could more than compensate for any biochemical costs due to the presence of a transgene. Hybrids derived from such transgenic crops could become invasive (Raybould and Gray, 1994). In a study conducted in the USA in 1996, transgenic thale cress, resistant to the herbicide chlorsulphuron, was twenty times more likely to transfer pollen to wild-types than an equivalent non-transgenic herbicide-resistant thale cress. This result suggested that the process of genetic modification itself might make it more likely that transgenes will spread to wild relatives, by somehow increasing the incidence of outbreeding, for example by affecting pollination or fertility (Bergelson *et al.*, 1998).

Crop–weed hybridization could therefore be ecologically disruptive. Technical measures are being developed to reduce the possibility of gene flow from transgenic crops (Chapter 8), but the simplest solution is to ban their cultivation in areas where the probability of genetic exchange with wild relatives is high. France has imposed a moratorium on the planting of transgenic oilseed rape cultivars for this reason. Canada has banned the growing of oilseed rape in the east of the country, where the probability of hybridization is high, while the cultivation of transgenic oats has been banned because weedy relatives are widespread (Brookes, 1998). The out-breeding crops most likely to transfer transgenes to wild weedy relatives include oilseed rape and canola, radish and other brassicas, rice, oats, sorghum and other cereals, cotton, sunflower, lettuce and artichoke (Keeler *et al.*, 1996). Transgenic crops need to be looked at case by case and assessed in terms of their potential for transferring transgenes. At the time of writing, many countries were only starting to take this into account when approving transgenic crops for cultivation.

Crop contamination: Impact on organic growers

Genetic contamination in crops can lead to agronomic and regulatory problems, but it is often the perception of contamination as much as the contamination itself that is damaging. This is the case for organic growers, whose products are sold to the public as being produced in an environmentally friendly manner, while being healthy and free from chemical contaminants such as pesticide residues. The public perception of organic food, at least in Europe, is that it is also free of contamination from transgenic crops.

The organization with the most influence in Britain when it comes to organic food is the Soil Association. Founded in 1946 to promote organic food and farming, it aims to represent consumers, food processors and farmers alike. Through its wholly owned subsidiary, Soil Association Certification Limited, it awards and administers the Soil Association Organic Symbol, used by around 75 per cent of organic producers and processors in Britain. Certification by the Soil Association is therefore crucial for organic farmers wanting to achieve a wide market. The Soil Association now issues licences to growers and producers only after establishing that there is no risk, as they see it, of organic produce being contaminated by transgenic crops. A slight possibility of genetic contamination can therefore adversely affect an organic grower.

Guy Watson's maize   An organic farmer in England, Guy Watson, backed by the Soil Association and Friends of the Earth (FOE) used legal means in 1998 to try and stop the flowering of a field trial of transgenic maize. The herbicide-resistant crop was grown on Hood Barton Farm, Dartington, situated about two kilometres from Watson's Riverford Farm, near Totnes, in Devon. Guy Watson specializes in growing organic produce for the expanding supermarket organic sector and for his own farm shop. The fear was that some of the pollen from the transgenic maize would fertilize his organic sweetcorn, threatening its organic status and considerably reducing its market value. The field trial was being run, on behalf of Sharpes International Seeds, by the National Institute of Agricultural Botany (NIAB) to compare the growth of four conventionally bred maize cultivars with a transgenic one. Sharpes argued that the like-lihood of pollen from the transgenic fodder maize contaminating the organic crop was too small to be measured (Masood, 1998).

Maize pollen can generally remain viable for anything from an hour to several days, depending on weather conditions. Male and female flowers are borne on the same plant, but the pollen is shed before the female flowers fully develop. Around five per cent of plants are self-pollinated, but the flowers of the remaining 95 per cent are fertilized by pollen from other plants. Maize is generally pollinated by wind, but bees find the flowers attractive and can also pollinate them. Maize produces large amounts of pollen, anything from 14 million to 50 million pollen grains per plant, enough to fertilize approximately 1,000 kernels. Within crop stands, around 50 per cent of the kernels of any individual maize plant are fertil-ized by pollen from plants growing within a radius of about twelve metres, with the other half being fertilized by more distant pollen sources. Maize pollen is one of the largest pollen grains dispersed by the wind. In low to moderate wind speeds, steep deposition gradients downwind of maize crops have been reported, while many pollen grains do not become effec-tively airborne in low wind speeds. Effective isolation distances of up to 400 metres have historically been recommended for maize, based on studies conducted in low to moderate wind conditions. Faster wind speeds result in dispersal of pollen over longer distances, although average viability may be reduced owing to impaction damage. Measurements of actual gene flow, using genetic markers to observe outcrossing, have suggested that pollen must on occasion travel longer distances. This probably occurs in conditions of gusty winds and cool temperatures, when pollen viability is high. In exceptional weather conditions, such as in storm fronts, pollen could be transported many kilometres. Honeybees can also transport pollen over distances of around four kilometres, although most cross-pollination by bees is done over much smaller distances (Emberlin *et al.*, 1999).

Guy Watson's case reached the High Court in July 1998, where the judgement hinged on the distance that viable pollen could travel. The judge ruled that the transgenic maize did not present a significant hazard to the organic maize and that it could be grown to harvest. This judgement was largely based on calculations made by the Government's Advisory Committee on Releases to the Environment (ACRE), who estimated that in the worst case, only one in 40,000 kernels of the farmer's crop would be fertilized by the transgenic pollen. This fell within the level of purity then required by international standards; the fields were judged to be ten times further apart than the minimum legal distance required to achieve this level of crop purity. The ACRE committee also noted that for cross-pollination to occur, both crops would have to flower simultaneously, while the sweetcorn would also have to be downwind of the transgenic maize (Masood, 1998). The government scientists agreed that a small proportion of pollen could travel long distances, in the order of at least one kilometre, and be long-lived in the environment, but the amount of cross-pollination at two kilometres was deemed too small to be significant. The judgement addressed the scientific probability of cross-pollination, but not the risk of the farmer's Soil Association accreditation being removed. Protesters destroyed the transgenic maize before it flowered and released its pollen (Gibbs, 1998). Since this case, organic farmers around the world have found their livelihoods increasingly threatened by the cultivation of nearby transgenic crops.

In October 1999, the European Commission's Standing Committee for Foods agreed on a legal threshold of 1 per cent for the accidental contamination of foods with GMOs. Industry and consumer groups largely welcomed this threshold, which gave food producers a standard to work with. The finding of more than 1 per cent contamination would compel manufacturers in European Union (EU) countries to label foods as 'containing GM material'. With less than 1 per cent contamination, however, manufacturers could not claim their products were 'GMO-free'. Many organic producers and suppliers in Europe, Australia and elsewhere were unhappy that 1 per cent transgenic material had become acceptable in food products. The Soil Association through its certification scheme is aiming for zero tolerance levels. This may be impossible to achieve in an environment where transgenic crops are being grown. Organic growers have warned that it may soon be impossible to grow organic vegetable crops in southern England, for example, if transgenic crops are grown commercially. Guy Watson has suggested that if transgenic maize seed is certified for growing throughout the south of England, 'every July and August, the air will be saturated with GM pollen. It will be impossible to grow an organic crop' (Meikle, 1998).

Fodder maize is one of the crops grown in the farm-scale evaluations in the UK (Chapter 11). These have continued to present problems to organic farmers. One field trial in 2001 was planned a little over three kilometres (two miles) from the Henry Doubleday Research Association at Ryton, near Coventry, Europe's largest research centre for organic crops. The centre carries out extensive trials on organic crops, including at least three maize cultivars, and is home to an important seed bank. The organic status of the research station (as certified by the Soil Association) was threatened, with serious consequences for its future operation. However, the transgenic maize trial was relocated in the face of strong protest, some of it from within the scientific steering committee advising on the trials. Nevertheless, around 31 industry-chosen farm-scale evaluation sites in the UK in 2001 (maize, oilseed rape and sugar beet) were within similar distances of organic farms (Brown and Lean, 2001).

Many other organic crops could soon be affected by the cultivation of transgenic cultivars. Pollen from transgenic cotton (*Gossypium hirsutum*) plants, for example, has been observed in neighbouring non-transgenic cotton. In a US study, only low levels of pollen occurred sporadically up to 25 metres from a cotton crop (Umbeck *et al.*, 1991). Pollen dispersal from a plot of transgenic cotton in Australia was also found to be relatively low, with an uncultivated buffer zone of twenty metres being adequate to limit dispersal of transgenic pollen from small-scale field trials (Llewellyn and Fitt, 1996). However, for large fields, growers of organic cotton adjacent to transgenic cotton may find it hard to avoid genetic contamination. A range of organic cereal and vegetable crops may also soon be affected, as the number of transgenic crop types being approved for commercial release increases.

Food contamination

The contamination of food through inadvertent gene flow from transgenic crops, particularly those that have not been approved for human consumption, is of increasing concern. In September 2000, for example, Aventis CropScience withdrew its StarLink transgenic maize in the USA after traces of it were found in taco shells produced with another maize cultivar. StarLink was engineered with a *Bt* gene (*Cry9C*) to protect the crop against the European corn borer, an important insect pest. The Environmental Protection Agency (EPA) had approved StarLink for animal consumption, but it was not approved for human consumption because the insecticidal *Bt* protein shares some properties with a known allergen.

Allergic reactions can be triggered by tiny amounts of an allergen in people who are sensitive to that type of chemical. Maize is therefore considered unfit for human consumption if one kernel in 2,400 contains the *Cry9C* protein. Concerns about the introduction of allergens to food made from transgenic crops first surfaced when a soyabean modified by Pioneer Hi-Bred, with a gene from Brazil nuts to enrich its protein content, caused allergic reactions in volunteers known to be sensitive to Brazil nut allergens (Nordlee *et al.*, 1996). However, in cases of accidental genetic contamination, the presence of allergens will be much harder to detect.

StarLink contamination was first found when FOE in the USA sent samples of Kraft taco shells, purchased from the fast-food chain Taco Bell, to an independent laboratory for testing. They were subsequently withdrawn from sale, along with around 800 other products in the USA that potentially contained StarLink contamination. Further contamination was subsequently found in other non-StarLink cultivars. It has now been estimated that around 430 million bushels of maize were contaminated in 1999, representing over four per cent of that year's total US maize production. The contaminated maize was recalled and used for animal feed and ethanol production. Contamination was also found in 2000, and up to 5 per cent of the total US maize crop for that year may also have to be withdrawn (Kleiner, 2000). By November 2000, StarLink contamination had also been found in corn starch products in Japan and South Korea.

It is not clear how the StarLink *Bt* gene initially got into the human food chain. It is probable that StarLink seed was mixed with other maize seed before being delivered to growers, but StarLink pollen could also have fertilized other maize cultivars growing nearby (Boyce, 2000a). StarLink contamination may prove very difficult to eradicate completely, because of cross-fertilization among maize cultivars in the field and the low threshold set for the *Bt* toxin. StarLink contamination is therefore a problem that may be around for some time to come. The economic consequences will continue to be serious. A range of food-processing businesses have been affected; for example, a Kellogg's cornflake factory had to be temporarily closed. The clean-up costs in the US alone for the 1999 contamination were estimated at more than US$1 million. Three executives at Aventis were sacked as a result, and the company set aside US$100 million to cover compensation for farmers and customers.

In the StarLink case, the transgenic maize was grown for animal feed. The consequences might have been more serious if, for instance, the transgenic crop was producing a biofuel that was toxic to humans. As cross-contamination appears difficult to avoid, and any transgene could find its way into the human food chain, environmentalists have argued that any genetically engineered food crop should first be proven safe to eat. However, many cultivars of maize, soya and other food crops are being developed to produce therapeutic proteins and vaccines, biofuels and a wide range of other industrial products. To ban these crops because they may contaminate human food appears a little extreme. Conventionally bred oilseed rape used to produce lubricant oil that is toxic to humans, for example, has been cultivated for a number of years. However, much more needs to be done to maintain crop integrity, including far stricter segregation procedures, while testing for genetic contamination must become more widespread and systematic.

Canola in Canada

The first field to be contaminated by the transfer of a herbicide-resistance transgene may have belonged to Canadian farmer Tony Huether. In 1997, he planted separate fields on his northern Alberta farm, each 30 metres apart, with canola (the type of oilseed rape grown in Canada) that resisted either Monsanto's Roundup®, Cyanamid's Pursuit®, or Aventis' Liberty® herbicides. In 1998, one of his canola fields was found to have volunteer plants surviving after two glyphosate (Roundup) sprays. The 57-hectare field had never been sown with glyphosate-resistant canola, but a nearby field had been planted in the previous year with a glyphosate-resistant Roundup Ready™ canola called Quest. It is believed that the transgene spread from one crop to another by pollen movement. In 1999, canola plants that resisted all three of the different herbicides used on the transgenic cultivars were found. These plants had therefore acquired genes from up to three different transgenic canola crops. They had become stacked in the genome of the weed oilseed rape and were all being expressed to confer herbicide-resistance. Huether was angry that no one had told him that transgenes could be so easily transferred. All the farmers growing Monsanto's transgenic oilseed rape had signed a Technical Use Agreement but the documentation mentioned nothing about possible transgene spread. To control the herbicide-resistant volunteer plants, more toxic herbicides, such as 2,4–D, have to be used. This contrasted with claims made by biotechnology companies that such toxic chemicals would become obsolete on farms where herbicide-resistant crops were grown (Leahy, 1998; Anon., 2000b).

By 1998, around 60 per cent of Canada's canola was grown from herbicide-resistant seed. Of this, a little under half was resistant to glyphosate (Monsanto's Roundup). The rest was resistant to glufosinate (e.g. Aventis' Liberty) or imidazolinone herbicides such as triazine (e.g. Cyanamid's Pursuit). Gene flow from herbicide-resistant canola to non-transgenic cultivars could soon be commonplace in Canada. Just five years after transgenic crops were introduced, the Canadian National Farmers' Union was lobbying the government to obtain compensation for unintended genetic contamination of crops due to transgene spread. Farmers, such as Stewart Wells of Saskatchewan, have found detectable genetic material from transgenic crops in produce and seeds from their non-transgenic crops. As with other organic growers, he finds that this genetic contamination represents a potential loss of earnings (Hoyle, 1999).

Monsanto own the patent to the Roundup Ready gene that confers resistance to its own Roundup herbicide. They insist that Roundup Ready seed is their property, under any circumstances, and are vigilant in policing its use to stop farmers saving seed for replanting or exchanging it between themselves. A Monsanto representative in Canada, Aaron Mitchell, has admitted that, 'we always expected that a level of natural outcross would occur within the species'. However, Monsanto placed the emphasis on farmers to sort the problem out by, 'talking to their neigh-bours about the canola they grow'. Farmers say it is highly inconvenient to keep track of what their neighbours are growing, and then planning their own strategies accordingly (Hoyle, 1999). The company insisted that problems had arisen only because farmers had not followed recommen-dations for treating volunteer plants in crops following oilseed rape (e.g. barley). This consists of spraying the herbicides 2,4–D or MCPA, which would have killed the volunteers that had acquired the Roundup Ready transgene. Monsanto now advise growers not to sow cultivars with different herbicide tolerances in the same or adjacent fields.

The stacking of different herbicide-resistance transgenes has also been demonstrated in field trials in Canada. In one study, pollen travelled over a twenty-metre roadway, resulting in the formation of hybrids with resis-tance to two herbicides (Downey, 1999). In addition, there have been many informal or unsubstantiated reports of 'genetic pollution' in canola from North America since 2000. Concerns about crop contamination in Canada were brought into focus during a court case involving Percy Schmeiser, which went to the heart of the conflict between the age-old right of farmers to replant seed and the intellectual property rights awarded to multinational corporations.

Percy Schmeiser's canola

When canola grown by Percy Schmeiser on his farm in Bruno, Saskatchewan, became contaminated with Roundup Ready seed, he found himself being sued by Monsanto. Schmeiser had grown canola for around 40 years, effectively developing his own varieties by replanting seed collected from each year's crop. He had purchased some pesticides from Monsanto in the past, but never seed. He decided to take a stand against the multinational company, unlike similarly accused farmers who have backed down and settled out of court. Monsanto took Schmeiser to court, accusing him of patent infringement for growing Roundup Ready canola without their permission and demanded financial damages, including 'technology fees' for growing their crop.

Schmeiser insisted that he planted his 1997 and 1998 crops with seed saved from his crops in the previous years, and that any transgenic Roundup Ready plants growing on his land must have been spread as seeds by the wind from neighbouring fields or from grain trucks travelling on roads adjacent to his farm. He counter-sued, accusing Monsanto of contaminating crops, libel, trespass, and disregard for the environment by introducing Roundup Ready into an area without proper controls. As Schmeiser had not bought Monsanto seed he had not signed the company's Roundup Ready contract giving them unrestricted access to farmers' fields (the contract also obliges farmers to buy fresh Monsanto seed each year and purchase only Monsanto brand-name herbicide). The Monsanto crop police, who also encourage farmers to snitch on their neighbours, came on to Schmeiser's land illegally, regardless (Anon., 2001b).

The verdict reached in a Canadian Federal Court in Toronto, in April 2001, was that Schmeiser had to pay Monsanto C\$15,450 (£6,930) compensation for the Roundup Ready plants that grew in his fields. Justice Andrew MacKay noted the high proportion of herbicide-resistant plants within some of Schmeiser's fields (around 95 per cent), and said it was the farmer's duty to destroy them once he had realized they contained the patented Monsanto transgenes. Schmeiser had sprayed a ditch with glyphosate against weeds, and noticed the herbicide-resistant crop plants. He seems to have particularly selected seed from these plants for replanting, although he claimed that he does not usually spray Roundup on his crops. The judge further ruled that it did not matter how the plants had initially got into his fields: they were Monsanto's property whether they had accidentally blown in or were deliberately planted.

This judgement in the favour of the multinational raises serious questions. It penalizes the farmer for Monsanto's inability to prevent genetic contamination. Monsanto say they will not demand compensation from farmers in cases of accidental appearance of Roundup resistance in fields of non-transgenic crops, but clearly intend to take action if they suspect that seed containing their patented transgenes has been replanted. However, farmers may wish to select and replant seed from those plants for other reasons. Schmeiser claimed that the contamination forced him to destroy a variety he had been developing through selective breeding for many years. This prompts the question of how farmers are supposed to know which plants within their fields contain patented transgenes, as they are indistinguishable from other plants (and not stamped 'property of Monsanto'). It also puts additional pressure on Canadian farmers to follow their neighbours and plant transgenic canola. It makes it increasingly difficult for farmers who want to grow conventional seed and select plants suited to the particular conditions found in their fields (landraces). They cannot stop transgenic seed or pollen entering their fields, via the wind, birds or insects, or on farm machinery, and they may be made to pay for it anyway (Anon., 2001b; Kleiner, 2001).

Oilseed rape in Europe   The approval for the commercial cultivation of transgenic herbicide-resistant oilseed rape in Europe has been delayed by environmental concerns (Gledhill and McGrath, 1997). The spread of transgenes has been one source of unease. A report published in 2000 by NIAB, for example, highlighted the fact that certain types of conventionally bred oilseed rape, called Varietal Associations, appear to be particularly susceptible to cross-pollination. This is because these varieties have a high proportion of male-sterile plants. Pollen from other varieties, including transgenic ones, is therefore more likely to fertilize the female flowers. Varietal Associations account for around eight per cent of the autumn oilseed rape grown in the UK. The accumulating data on gene flow from transgenic oilseed rape suggests that the transfer of transgenes to conventional crops and wild relatives is inevitable if they were to be grown on any scale in Europe.

The spread of herbicide-resistance transgenes to a non-transgenic crop was first reported in Europe in the summer of 2000: in sugar beet. In experimental trials, transgenic sugar beet engineered for resistance to glufosinate was found to have accidentally acquired the genes to resist glyphosate. When glyphosate was sprayed onto the plots to kill the plants at the end of the experiment, many plants survived. This occurred in experimental plots in Britain, France, and The Netherlands. In each case, up to 0.5 per cent of sugar beet plants survived being sprayed with a herbicide to which they should have been susceptible, which represents a sizeable number of plants. In this case, the contaminating transgenes were transferred to the sugar beet in greenhouses operated by Aventis. Sugar-beet plants used to create the seed for the field experiments were fertilized by pollen from another transgenic variety, engineered to resist a different herbicide (MacKenzie, 2000).

As we have previously noted, herbicide-resistant crops need to be sus-ceptible to at least one group of herbicides in order to control them if they become weeds. European regulations, for example, forbid the creation of plants resistant to several herbicides for this reason. Herbicide-resistant crops that pick up additional herbicide-resistance transgenes in the field become doubly resistant. More noxious and environmentally damaging herbicides have to be used to combat these plants. In Canada, plants have acquired resistance to three different herbicide groups via transgene spread and gene stacking. There is now a growing body of evidence to suggest that the widespread cultivation of herbicide-resistant crops could lead to the creation of volunteer weeds with stacked transgenes. These plants are likely to be the first 'superweeds' created by the cultivation of transgenic crops.

Seed contamination: The Advanta case

The contamination of seed due to gene flow from transgenic crops is becoming an ever bigger issue, especially in Europe. In spring 2000, for example, after being given government reassurances that the commercial growing of transgenic crops would be delayed until a series of farm-scale evaluations had been completed, the British public learned that conven-tional oilseed rape had been widely contaminated with transgenic material. The oilseed rape in question may have been used to make food-stuffs such as cooking oil, margarine, and ice cream, in addition to being used as animal feed. The imported seed, supplied by Advanta Seeds (at this time part-owned by AstraZeneca and Royal Vandehave), came from Alberta, in the Canadian prairies, from seed crops that had been grown more than 800 metres from the nearest transgenic crop. They had, nevertheless, picked up transgenes. The seed was contaminated by cross-fertilization with pollen from transgenic cultivars growing up to half a mile away. The company was not aware of any problems, although it did no seed testing for possible genetic contamination, despite the growing evidence that such contamination was likely.

The unwanted transgenic material in the seeds was discovered in Germany during purity testing of imported seeds by the state of Baden-Wurttemberg in late March 2000 (a system of testing not done in Britain at that time). Advanta Seeds was immediately told, and informed its British division in early April. The company took steps to prevent further sale of the seeds and did its own testing, which revealed up to 1 per cent transgenic material in seed produced in 1998, which was sold over the following two years. Seed produced from 1999 onwards was not contaminated, because seed crops were moved to new areas: Ontario, Canada; Montana, USA; and New Zealand. These are all areas where cross-pollination with transgenic crops is much less likely. The company told the British government, and key advisers (English Nature) and the public were informed of the situation a month later (Lean, 2000; Meikle, 2000a).

Around 400 British farmers planted the Advanta oilseed rape, in the spring of 1999 and 2000. The contaminated seed was grown on a maximum of 9,000 hectares in 1999, amounting to nearly two per cent of the total oilseed rape crop. The seed was also grown on around 600 hectares in France, 500 in Sweden, and 400 in Germany (Meikle, 2000a). Farmers in Britain were alarmed to learn in late May 2000 that major supermarket chains, including Tesco, Safeway and Asda, were not touching products from farms that had grown the contaminated seed. This was followed by an announcement by the Seed Crushers' and Oil Producers' Association (Scopa), which processes harvested plants into the oil used in foodstuffs, that they would not accept contaminated oilseed rape. This rendered the crop effectively worthless. The contaminated crops were all eventually ploughed in and destroyed (Arthur, 2000). In June 2000, a compensation package was offered to farmers. Advanta Seeds, under political pressure and admitting no liability, paid affected farmers a total of around UK£1.5 million (Meikle, 2000b). In the wake of the Advanta incident, the British environment minister Michael Meacher conceded that there was no way of preventing transgenic crops contaminating neighbouring conventional crops (Meikle, 2000c).

The Advanta case revealed the need for more stringent checks on seed purity now that transgenic crops are being grown on a large scale. However, the Advanta case is not an isolated one. In a survey of twenty randomly chosen samples of conventional seed from American distributors, conducted by Genetic ID of Iowa in December 2000, half were found to have traces of contamination from transgenic cultivars (LePage, 2000). This is the result of sloppy practices by major seed companies.

In Europe it has been left to individual governments to deal with genetic contamination of crops. In March 2001, Italian police seized 400 tons of soyabean seed and 130 tons of maize seed because they suspected that it might contain significant amounts of genetically modified material. Cultivation of transgenic crops is banned in Italy. These claims of genetic contamination subsequently proved unfounded. However, the EU Scientific Committee on Plants has said that the presence of GM material in seeds imported from the USA is now inevitable because of unintentional contamination in the production process. The EU has now outlined draft legislation for dealing with the presence of GM contamination and GM seeds in conventional cultivars, in Directive 90/220 covering deliberate releases of GMOs into the environment. This specifies that contamination should not exceed a threshold of 0.3 per cent in cases of cross-pollinated varieties and 0.5 per cent in cases of self-pollinating and vegetatively propagated crops, while there remains zero tolerance for GM seed contamination not authorized under Directive 90/220 (Dorey, 2001).

Advanta moved its oilseed rape crops to new areas when it became aware of its genetic contamination problem. However, there may come a time when all areas in Canada and the USA, for example, will be growing transgenic crops. Will a certain level of genetic contamination become acceptable in all crops grown for their seed? As more countries grow transgenic crops over an increasing area, it may soon become impossible to produce truly transgene-free seed or to grow 'GMO-free' agricultural produce.

7 Impact on non-target species

Transgenic crops have been modified with the aim of impacting on only one or a few target pest or disease species or, in the case of herbicide-resistance, only the crop plants themselves. This is fine in theory, but species are closely interconnected within communities in field situations. In this chapter, evidence for a range of potentially adverse indirect impacts on non-target species is reviewed, including natural enemies of pests, wildlife such as butterflies and farmland birds, and pollinating insects.

Natural enemies

A natural enemy is a beneficial species within an agricultural ecosystem (agroecosystem), which naturally feeds on or parasitizes a pest species. An abundance of natural enemies will reduce the economic impact of a pest species, but any depletion of natural enemy populations may cause a resurgence in crop pest attacks. It has been said that when natural enemies are killed, we inherit their work. Pesticides that harm natural enemies, for instance, create demand for more pesticides (the pesticide treadmill). A number of unforeseen impacts of transgenic plants on natural enemy species have been reported.

A number of transgenic plants have been engineered in experimental programmes with the aim of increasing plant resistance to aphids, commonly referred to as greenfly or blackfly, which are major insect pests of a wide range of crops. In the field, the natural enemies of aphids include three important groups of voracious predators, ladybird beetles (coccinellids), lacewings (chrysopids), and hoverflies (syrphids); and a group of parasitic wasps or parasitoids. Adult and larval ladybirds both consume aphids, while only the larval stages of lacewings and hoverflies are predatory. Ideally, for pest management purposes, insect-resistance in plants should be compatible with other components of pest control, including biological control by natural enemies.

Ladybirds In one research programme, a gene expressing a type of protein called a lectin, isolated from snowdrops (*Galanthus nivalis*), was incorporated into potatoes to stop aphids from feeding on them (Coghlan,

1996). Lectins bind specifically to sugar molecules, particularly those present on the surface of cells, causing them to become 'sticky'. It is the ability of lectins to cause cells to agglutinate or stick together that makes them toxic to herbivorous insects. Similarly, a different type of lectin in uncooked kidney beans renders them toxic to humans. No lectin-containing transgenic crop has to date been approved for commercial release. Nick Birch and colleagues at the Scottish Crop Research Institute (SCRI) reported adverse effects on adult 2-spot ladybirds (*Adalia bipunctata*), feeding for twelve days on peach-potato aphids (*Myzus persicae*) reared on transgenic potatoes engineered with a lectin gene. In a laboratory study, female ladybirds that consumed aphids reared on transgenic potatoes lived half as long as those feeding on aphids from conventional potatoes. In addition to the 51 per cent reduction in longevity, females eating aphids from transgenic potatoes laid 30 per cent fewer viable eggs. This adverse effect on ladybirds was therefore transmitted via the food chain. The effect was temporary, and if the ladybirds that had consumed aphids reared on transgenic plants were switched to a diet of aphids reared on non-transgenic plants, the effect was reversed (Birch *et al.*, 1999). Conversely, another study showed no significant effects on the development of the convergent ladybird (*Hippodamia convergens*) as a result of them feeding on peach-potato aphids reared on transgenic potatoes containing a *Bt* transgene. However, it was unclear to what extent the toxin was present in the aphids consumed by the ladybirds (Dogan *et al.*, 1996).

Tri-trophic interactions   Feeding relationships within ecosystems can be seen in terms of stages in a food chain, called trophic levels. During the 1980s, ecologists realised that to understand insect–plant interactions in greater depth a third trophic level, comprising the predators and parasites of the plant-consuming insects, needed to be taken into consideration. Similarly, the plant trophic level needed to be taken into account when studying interactions between herbivorous insects and their natural enemies (Barbosa and Letourneau, 1988). Plants being attacked by herbivores, for example, were shown to release volatiles that attracted natural enemies (parasitic wasps) specific to those plant-feeders (Turlings *et al.*, 1990; Tumlinson *et al.*, 1992).

The study of tri-trophic interactions has revealed many complex and subtle interactions between organisms occupying different trophic levels. The SCRI's potato–aphid–ladybird study was the first to demonstrate conclusively a tri-trophic interaction involving a transgenic plant, a plant-consuming insect, and a beneficial natural enemy of that insect. The study of tri-trophic interactions is particularly important if transgenic crops are to be successfully used in Integrated Pest Management (IPM) systems. If an insect-resistant transgenic crop also adversely affects that pest insect's natural enemies, then the agricultural benefits of insect-resistance will be diminished.

Lacewings A number of tri-trophic effects involving transgenic crops and non-target beneficial species have now been reported. In one of these, conducted in a laboratory in Zurich, Switzerland, by Angelika Hilbeck and her co-workers, predatory lacewing (*Chrysoperla carnea*) larvae were fed on herbivorous European corn borer (*Ostrinia nubilalis*) and armyworm (*Spodoptera littoralis*) caterpillars that had been reared on transgenic *Bt* maize (corn). The maize expressed a *Cry1A* gene, which produces a *Bt* toxin that is highly toxic to European corn borer and caterpillars of several other moth and butterfly (Lepidoptera) species. Lacewing larvae feeding on both prey species reared on *Bt* maize had a higher mortality than larvae fed on prey raised on conventional plants (62 per cent compared to 37 per cent mortality). Development time was also prolonged when lacewings were fed on prey reared on *Bt* maize. The authors concluded that the combined effect of *Bt* exposure and nutritional deficiency caused by consuming sick prey was responsible for the adverse effect on lacewings (Hilbeck *et al.*, 1998).

Parasitic wasps Adult female parasitoids locate their insect host, which is different for different parasitoid species, in order to lay an egg in it. The immature stages of the parasitoid develop inside the host insect. As we have noted, parasitoids can use volatile chemicals released by plants when insects attack them to help them locate their insect hosts. This is an example of a chemically mediated interaction. Chemicals that mediate interactions between organisms are called semiochemicals. Pheromones are semiochemicals that mediate intraspecific interactions; for example, sex pheromones emitted by female moths attract males of the same species. Allelochemicals are semiochemicals that mediate interactions between different species. They have been classified in terms of the species they benefit. A kairomone benefits the organism receiving the chemical signal; an allomone benefits the emitter of the signal; while a synomone benefits both the receiver and emitter. The volatiles released by plants that attract parasitic wasps are therefore synomones; their production benefits the plant and the parasitoid (but not the intervening herbivore) (Nordlund and Lewis, 1976).

The wasp *Cotesia plutellae* is a parasite of the diamondback moth (*Plutella xylostella*), an important pest of brassica crops, including cabbage. This herbivore–parasitoid relationship has become a model system in the investigation of tri-trophic interactions. In a wind tunnel study, conducted at Rothamsted Experimental Station, England, by Tanja Schuler, Guy Poppy and their colleagues, diamondback moth caterpillars caused less feeding damage on *Bt* oilseed rape leaves than on wild-type leaves. The transgenic oilseed rape contained a *Cry1A* gene, producing a *Bt* toxin that is particularly active against moth pests. Therefore, the reduced feeding damage was expected. The parasitic wasp was then released downwind of transgenic and conventional oilseed rape, and landing behaviour on the leaves was recorded. Less-damaged leaves were chosen as landing sites less often by parasitic wasps. The *Bt* oilseed rape leaves were therefore effectively avoided, most likely due to fewer attractant volatiles being released from the less-damaged leaves. When moths with resistance to *Bt* were used in the experiments, however, they consumed as much of the *Bt* leaves as the conventional leaves, and parasitoids located them as often on both leaf types (Schuler *et al.*, 1999). In this case, the tri-trophic interaction may be to the ultimate advantage of the crop, as the parasitoid may preferentially locate *Bt*-resistant host insects, and thereby help to constrain the spread of *Bt*-resistance genes within the pest population.

These studies demonstrate how transgenic crops can affect tri-trophic interactions. Much more research needs to be done in this area in order to understand the subtle ecological impacts of transgenic crops on non-target species. These impacts might have implications for pest management, by adversely affecting beneficial species, or for conservation, by affecting vulnerable native species.

Monarch butterfly

The monarch butterfly (*Danaus plexippus*) is a large, brightly coloured migratory species that travels south, to Florida, California and Mexico, to overwinter and then north, to the mid-western states of the USA and Canada, to breed during the summer. Monarch caterpillars feed exclusively on milkweed plants (*Asclepias*), from which they sequester toxins (cardiac glycosides) into their bodies to deter predators. The butterfly is a flagship conservation species, and its limited host-plant range makes it vulnerable (Malcolm *et al.*, 1993). It was only after transgenic *Bt* maize was being grown on an ever-increasing proportion of the North American Corn Belt, however, that ecologists first started to look at possible impacts on monarch butterfly populations.

The *Bt* toxin is expressed in the pollen of commercially grown transgenic *Bt* maize, although it is only in the stems or leaves that target pest insects encounter it. Maize pollen can be dispersed at least 60 metres by wind, and can come to rest on other plants. Pollen from transgenic maize, expressing a *Cry1A* gene, had an adverse effect on monarch butterfly caterpillars, in a laboratory study conducted at Cornell University, USA, by John Losey and his colleagues. Pollen from a Novartis transgenic maize (N4640–*Bt*) and a conventional maize were collected and dusted on to leaves of milkweed plants (*Asclepias curassavica*). The density of pollen was set to visually match that of milkweed leaves observed in maize fields. Individual leaves were taken from these plants and placed in water-filled tubes, and five laboratory-reared monarch caterpillars were placed on each leaf. The caterpillars consuming *Bt* pollen ate less, grew more slowly, and suffered higher mortality than caterpillars reared on leaves dusted with conventional maize pollen or control leaves with no pollen. After four days, only 56 per cent of caterpillars remained alive on leaves dusted with *Bt* pollen, compared to 100 per cent on the other treatments (Losey *et al.*, 1999).

This finding was supported by a study from Iowa, conducted by Laura Hansen Jesse and John Obrycki, in which leaves from common milkweed plants (*Asclepias syriaca*) adjacent to transgenic maize fields were bought into the laboratory and fed to monarch caterpillars. After 48 hours, 20 per cent mortality had occurred on the leaves naturally dusted with *Bt* pollen, compared to 3 per cent mortality with pollen from non-transgenic maize, and zero mortality on pollen-free leaves. It was concluded that the effects of transgenic pollen may be observed at least ten metres from field borders, with highest monarch caterpillar mortality on milkweed plants likely to occur in transgenic maize fields or within three metres of their edge (Hansen Jesse and Obrycki, 2000).

The results of these studies have potentially serious implications for the conservation of monarch butterflies. Transgenic maize may soon make up a third of the total US maize crop, while milkweed plants typically occur around maize fields. A large proportion of monarch caterpillars could be in or near transgenic maize fields during pollination. Maize sheds its pollen between June and mid-August, during the time when monarch caterpillars are feeding. Although the northern range of the monarch is large, 50 per cent of its summer range is within maize-growing areas. Widespread *Bt* maize cultivation could therefore have a significant impact on monarch populations (Losey *et al.*, 1999).

The preliminary results from several studies commissioned by the biotechnology industry, in response to concerns raised by the Cornell study, suggested that pollen containing *Bt* toxin may not present as large a risk as the laboratory study might suggest. Different strains of *Bt* in trans-

genic maize are known to have differing toxicity, although the Cornell
study involved a strain with a relatively low insecticidal toxicity. In the
field, pollen is much more likely to be blown or washed off leaves than in
the laboratory. Adult monarchs may avoid laying eggs on milkweeds next
to maize plants, and on leaves with heavy coatings of pollen. Pollen concentrations may drop off rapidly several metres from maize fields, suggesting that milkweeds further from maize fields will be relatively
pollen-free. Meanwhile, in Nebraska, 95 per cent of maize pollination was
shown to be complete before the first monarch eggs on milkweed hatched.
However, many of the scientists present at a meeting in Chicago in 1999,
where these results were discussed, argued for a more cautious approach
to the planting of transgenic maize (Niiler, 1999; King, 2000).

The monarch studies should be placed within a wider context. More
than 80 million lb of conventional pesticides were applied to US crops in
1997, according to the Environmental Protection Agency (EPA, 1998).
Widespread effects on non-target organisms result from conventional pesticides. Maize has received more pesticide treatments than almost any
other US food crop in recent years, but still large yield losses have
occurred. The European corn borer is particularly difficult to control by
spraying, because it burrows inside the plants. Plants with inbuilt *Bt* toxins,
therefore, represent a more effective control strategy. It has been estimated
that the growing of transgenic *Bt* maize, in conjunction with rotations and
pest monitoring, could reduce insecticide use by more than 50 per cent
(Pimentel and Raven, 2000). To date, however, little reduction in pesticide
use has occurred in *Bt* maize, because farmers have been using the crop's
in-built insecticidal properties as extra 'insurance', while continuing to
spray insecticide as before.

Monarch numbers have declined during the 1990s, with widespread
insecticide applications playing a part in this, although habitat destruction
is probably the main factor. The monarch's overwintering sites in Mexico,
where butterflies congregate, are particularly vulnerable (Lewis and
Palevitz, 1999). Monarch butterfly populations declined in Mexico from
170–204 million to only 56–67 million, between 1996–97 and 1998–99,
according to figures presented by Monsanto as part of its defence of *Bt*
maize (Monsanto, 1999). In addition, many farmers view milkweed as a
toxic plant, fit only for weeding out. Milkweed has become more common
in maize fields, as well as in soyabean and wheat rotations, in recent times
owing to the use of more selective herbicides and reduced mechanical control of weeds. The presence of milkweed in a crop can cause significant
yield reductions (Yenish *et al.*, 1997).

The importance of the monarch butterfly as a conservation symbol, and the widespread coverage afforded the original Cornell study, contributed to a more sceptical public attitude to transgenic crops in the USA. A number of scientists were quick to criticize the *Bt* maize pollen work for this reason (Shelton and Roush, 1999). Nevertheless, significant monarch mortality has been observed with pollen levels found in the field, over just 48 hours of feeding. The findings have acted as a timely 'warning bell', according to Linda Rayor, one of the scientists involved in the Cornell study. Further work on the ecological impacts of insecticidal crops is clearly needed, including an assessment of the effects of maize cultivars expressing different *Bt* toxins on non-target species (Radford, 1999).

Black swallowtail   Research groups have now started to look at other non-target butterfly species potentially at risk from pollen containing *Bt* toxin. May Berenbaum's group at the University of Illinois at Urbana, USA, for example, studied the effects of *Bt* pollen on the black swallowtail (*Papilio polyxenes*) in both the laboratory and field. Black swallowtail caterpillars feed almost entirely on apiaceous species, such as wild carrot and wild parsnip, that in the Midwestern USA are located in the vicinity of maize fields.

In the Urbana group's laboratory study, pollen was collected from *Bt* maize (Pioneer's 34R07 and Novartis' Max 454 varieties), and a non-*Bt* variety (Pioneer 3489). Leaf disks were cut from greenhouse-grown wild parsnip, and a range of pollen concentrations were placed on to them. Single young caterpillars were placed on each leaf disc. The number of caterpillars surviving after three days was recorded. There was no difference in mortality for any of the pollen concentrations between the Pioneer transgenic and non-transgenic cultivars. However, there was a significant effect for the Novartis Max 454 pollen at the highest concentration compared to all the other treatments; this was 40 times higher than the highest pollen exposure recorded in the accompanying field study (Wraight *et al.*, 2000).

The Urbana group's field experiments were conducted along the edge of a field planted with Pioneer transgenic crop (with pollen expressing a *Cry1A Bt* gene). Rows of potted wild parsnip plants were spaced at increasing distances from the field edge (0.5–7 m from the crop), and ten young black swallowtail caterpillars were placed on each plant. Pollen falling on the plants was monitored using a wire-staked microscope slide covered with a thin layer of Vaseline. The weight and percentage mortality of caterpillars on the plants was recorded for a week. The plants were downwind of the crop for 43 and 71 per cent of the days, on the two occasions that this experiment was conducted. Pollen load on leaves declined from 210 and 100 grains/cm² at 0.5 m, to 26 and 11 grains/cm² at 7 m from the crop, for these two experiments. Caterpillar mortality was high, probably due to predation. No significant relationships between larval survivorship or weight were detected either for distance from the field edge or for pollen level. Caterpillars feeding less than one metre from the field were therefore as healthy as those feeding seven metres away, despite ingesting five times as much of the pollen (Wraight *et al.*, 2000). One line of *Bt* maize (Syngenta's KnockOut) has since proved harmful to both monarch and black swallowtail caterpillars in field studies.

A range of beneficial insects is also potentially at risk from the toxins in *Bt* pollen. A ladybird (*Coleomegilla maculata*), a lacewing (*Chrysoperla carnea*) and an anthocorid bug (*Orius insidiosus*) were found to be unaffected when fed *Bt* maize pollen in one study (Pilcher *et al.*, 1997). More studies of this type are needed, however, as the effects of *Bt* pollen may present significant risks to non-target insects, not just with maize but also for other transgenic crops expressing *Bt* toxins.

Farmland birds

The cultivation of herbicide-resistant transgenic crops could adversely affect farmland birds, which feed on the seeds of weeds and the insects that live on weeds, if such cultivation leads to increased spraying of herbicides. In Britain this has emerged as a particularly important concern. English Nature, the government's nature conservation advisory body, called for a three-year moratorium on the commercial growing of transgenic crops in a 1998 report, which cited the potential impact on native bird populations as one of the main reasons. The report was backed by a number of Britain's largest non-governmental organizations, including the Royal Society for the Protection of Birds (RSPB) and Friends of the Earth (FOE) (Vidal, 1998a). In Britain, nature and agriculture exist in close proximity, in contrast to, say, the USA, where agriculture (e.g. the Corn Belt) and wildlife (National Parks and wilderness areas) tend to be more compartmentalized. In Britain, around 76 per cent of land use is agricultural, while in the USA the figure is 35 per cent. Therefore, in Britain and many other European countries, farmland is synonymous with the countryside. Changing farm practices can therefore significantly affect wildlife.

Populations of many farmland birds are declining all over Europe. In the British Isles between the early 1970s and the late 1990s, for example, populations of corn bunting (*Emberiza calandra*) declined by 74 per cent, partridge (*Perdix perdix*) by 78 per cent, tree sparrow (*Passer montanus*) by 78 per cent, skylark (*Alauda arvensis*) by 60 per cent, yellowhammer (*Emberiza citrinella*) by 60 per cent, and the linnet (*Acanthis cannabina*) by around 50 per cent. Modern farming methods have brought about these dramatic declines. Important factors include the removal of hedgerows, increased use of herbicides and other pesticides, and the planting of dense winter-sown crops (Avery and Gibbons, 1999; Campbell and Cook, 1995).

Herbicide-resistant crops could potentially exacerbate bird population declines, because weed control can be more thorough in crops that tolerate high levels of herbicide. This can directly reduce the seed supply available to feeding birds, while indirectly affecting those bird species that feed on insects and other invertebrates that live on weeds in fields, field margins and hedgerows. The programme of farm-scale evaluations being conducted in Britain was set up to investigate the indirect ecological impacts of herbicide-resistant crops (Chapter 11). If the results of these trials show adverse ecological effects, the British government will be under intense pressure from environmental groups to ban the growing of herbicide-resistant transgenic crops permanently.

One problem relating to farm-scale trials in general is that birds range over many kilometres in search of food. In the British farm-scale evaluations, rather than bird numbers, biodiversity indicator species are quantified, including weed and insect species that form the diets of many farmland birds. Alternatively, mathematical models can be used to predict possible impact on a regional scale. A computer modelling study, conducted by Andrew Watkinson and colleagues at the University of East Anglia in England, predicted how the introduction of herbicide-resistant sugar beet would affect the population dynamics of fat hen (*Chenopodium album*), an annual weed that is an important source of food for farmland birds. The study showed how the planting of transgenic herbicide-resistant sugar beet could affect seed-eating bird species, such as the skylark, via its impact on weed populations. Skylarks depend on the seeds from common weeds like fat hen, especially in the autumn and winter when alternative food sources are scarce. Therefore, the greater the density of weeds in an area, the more skylarks can be supported. The computer modellers suggested that the introduction of transgenic crops could reduce the number of weeds in fields of sugar beet by more than 90 per cent. The model predicted that if all farmers who currently have difficulty in controlling weeds planted herbicide-resistant sugar beet, then the number of skylarks could fall by an additional 80 per cent (Watkinson *et al.*, 2000; Radford, 2000).

The East Anglian model could be adapted for any situation involving weeds, seed-eating birds and a technological innovation. Such research makes it clear that if farmland birds are to be conserved, a balance has to be struck between effective weed control and biodiversity considerations. If spraying was limited to occasions when weeds exceeded a certain density (threshold), for example, then the impact on skylarks was shown to be less severe (Watkinson *et al.*, 2000; Edwards, 2000). Another option may be to grow herbicide-resistant crops in combination with unsprayed headlands (the ends of fields, where ploughs historically turned) or field margins (strips of land between the crop and field boundary, extending a limited distance into the crop), where the weeds that support bird populations can thrive. A compromise along these lines is available to the British government if it wants to back industry and give the go-ahead to transgenic crops, but the farm-scale evaluations show adverse impacts on biodiversity.

Bt toxins in the soil

Insecticidal toxins from *Bacillus thuringiensis* (*Bt*) are known to remain active in soils, particularly clay and humic acid soils, to which they can bind. Laboratory experiments with toxins from *Bt* subspecies *kurstaki* (*Btk*), which is toxic to larvae of butterfly and moth caterpillars, and *Bt* subspecies *tenebrionis* (*Btt*), which is toxic to beetle larvae, have been done to quantify the amount of toxin that binds to soil particles. Standard amounts of clay minerals adsorbed the *Bt* toxins rapidly (within 30 minutes in the case of *Btk*). The amount of toxin adsorbed increased with concentration, until a plateau was reached. Repeated washes dislodged only 10–30 per cent of the toxins from the clay particles; the rest was bound tightly in a clay–toxin complex (Tapp *et al.*, 1994). The *Bt* toxin–clay complexes resisted degradation, in contrast to free *Bt* toxin in soil, which is susceptible to relatively rapid microbial degradation (Crecchio and Stotzky, 1998).

The bound *Bt* toxin retained its insecticidal activity, which varied with the soil type to which it was bound. Insecticidal activity was tested using tobacco hornworm (*Manduca sexta*) caterpillars, which are model insects frequently used in laboratory studies. Soil suspensions, with and without bound toxin, were distributed on the surface of the insect's regular diet. The concentration of bound toxin that caused 50 per cent mortality to young caterpillars after seven days was calculated. In the case of a clay soil type called kitchawan, the bound *Btk* toxin remained toxic for around six months (Tapp and Stotzky, 1998).

In another series of experiments, seeds of transgenic maize containing a *Bt* toxin gene were grown in sterilized soil. Extracts of soil were analysed for the presence of *Bt* protein, and subsequent insecticidal activity. Toxin from transgenic *Bt* maize was shown to persist in the soil extracted from around plant roots. It bound to soil particles and remained toxic to tobacco hornworm caterpillars for many weeks (Saxena *et al.*, 1999).

In *Bt* bacteria naturally occurring in soil, the toxin is initially produced as an inactive precursor, which is later converted to an active form. However, in transgenic crops modified with the *Btk* toxin, a truncated *Cry1A* gene produces the active *Bt* toxin continuously in most green tissue. In 1998, around 20 per cent of the total maize crop in the USA was grown from *Bt* maize. In 1999, this increased to around 30 per cent. As the cultivation of this transgenic crop becomes more widespread, more *Bt* toxin is likely to enter the soil, in a form that could be persistent. This may pose a hazard to non-target organisms (Saxena *et al.*, 1999).

Pollinating insects and honey contamination

Pollen may have an impact on the insects that visit flowering transgenic crops. Pollen containing novel proteins might prove toxic to pollinating insects, or influence their physiology or behaviour in adverse ways. In the case of honeybees this might affect their ability to pollinate fruit crops or to produce honey. Hives are often imported into orchards to ensure that pollination occurs and fruit develops. Bees are important pollinators also in a range of field crops. Adverse impacts on honeybees therefore could have significant economic impacts. There is also the possibility that novel proteins expressed in the pollen of transgenic crops might find their way into honey.

An experimental transgenic oilseed rape modified for resistance to a fungal disease, using a transgene expressing an enzyme called chitinase, increased the plant's tolerance of fungal infection. The chitinase breaks down a chemical called chitin, which gives fungi their structure. However, concern was expressed that the enzyme might do physiological harm to honeybees, because chitin is a ubiquitous component of insect cuticle (the hard outer coat). Experiments with chitinase and bees showed the initial fears in this case to be unfounded, although the crop in question has never been commercialized (Pham-Delegue *et al.*, 1992).

The possibility that transgenic crops might affect the foraging behaviour of honeybees was raised during a three-year collaborative project, involving researchers in France, Belgium and Britain, that started in late 1996 to look at bee pollination of transgenic oilseed rape. A non-commercial oilseed rape that expressed an insecticidal proteinase inhibitor had levels of toxin that were too low to detect in the pollen or nectar, although it is possible that the toxin could become more concentrated within beehives. Pollen is a natural source of protein for honeybees, while honey is also a food source. In experiments, honeybees were fed high levels (up to 100 times that in oilseed rape tissue) of proteinase inhibitors in sugar solutions. Bees that had fed on this solution for three months died fifteen days earlier, on average, than bees fed on a normal sugar solution. Bees fed on sugar solutions containing cowpea trypsin inhibitor (CpTI), a proteinase inhibitor engineered into modified oilseed rape, were less efficient at learning to distinguish between the smells of different flowers than bees fed on regular sugar solution. Although there are currently no plans to release commercially oilseed rape lines containing proteinase inhibitors, this study demonstrated how the presence of transgene products in nectar or pollen could affect the efficiency of foraging behaviour in pollinating insects (Crabb, 1997; Picard-Nizou *et al.*, 1997).

The possible contamination of honey with pollen from transgenic crops has been of increasing concern. *Bt* toxins, for example, could remain bioactive in honey for several weeks. Although they are toxic only to insects, and not to humans or any other mammal, they could act as allergens. Any transgenic material from pollen is likely to be present only in very small amounts, but even trace amounts of novel proteins can potentially have allergenic effects. Consumers regard honey as a pure and natural product, and the mere 'taint' of genetic engineering has the potential to adversely affect sales. In Britain, FOE has initiated research to try and establish the extent to which honey could become contaminated.

In 1998, FOE showed that pollen from transgenic oilseed rape was carried for more than 2.5 miles (about 4 kilometres) by foraging honeybees. They subsequently contracted the National Pollen Research Unit at University College, Worcester, and independent bee and honey consultants, to monitor pollen around a farm-scale evaluation of spring oilseed rape in Oxfordshire (Model Farm, near Watlington) in summer 1999. Airborne pollen was collected from around the sites and pollen-traps were sited at six beehives, two at 500 metres, two at 2.75 kilometres, and two at 4.5 kilometres from the crop. The pollen samples were sent by FOE and 'Newsnight' (a BBC TV current affairs programme) to the laboratory of the Federal Environmental Agency in Vienna, Austria for DNA analysis. Airborne pollen from the transgenic crops was found up to 475 metres from the trial site, while all the beehives monitored were found to contain transgenic pollen (FOE, 1999a). The finding of pollen 4.5 kilometres from oilseed rape is consistent with previous reports of pollen movement over distances of 4 kilometres or more (Chapter 6). This represents a potentially serious contamination problem for honey producers.

Traces of pollen with identifiable transgenes were also found in locally produced honey bought in a shop in Oxfordshire, not far from the farm-scale evaluation of transgenic oilseed rape. Two samples, analysed by the independent Vienna laboratory, showed traces of DNA indicative of trans-genes specific to the herbicide-resistant crop developed by Aventis and used in the trials (FOE, 2000). The non-target species potentially affected by this and other cases of genetic contamination of food, such as the tacos discussed in the previous chapter, is, of course, humans. The levels in honey were well below the 1 per cent threshold allowed for accidental genetic contamination set by the European Union (EU) and risks to human health were remote. Nevertheless, transgenic crops involved in the farm-scale evaluations had not been cleared for human consumption in Britain.

The farm-scale evaluations in Britain became of great concern to the members of the Bee Farmers Association, 350 of whom were large-scale (over 40 hives) bee farmers whose livelihood depended on their bees. The Association advised its members to move their hives at least six miles (9.7 km) away from the nearest transgenic crop trial. Many bee farmers derive a large part of their income from supplying bees for the pollination of orchard and vegetable crops. They provide a vital pollination service to growers of top fruit (apples, pears, plums and cherry) and soft fruit (straw-berries and raspberries). The value of pollination services in Britain is estimated to be £200 million annually. This use of honeybees has grown in importance due to a parasitic mite (*Varroa*) precipitating a decline in native bee populations. Both bee farmers and fruit growers were therefore financially affected by the local planting of transgenic crops (FOE, 2000; McCarthy, 2000).

Studies of honeybee foraging behaviour have confirmed that pollen from transgenic oilseed rape is likely to be widely collected by bees. Honeybees are the primary pollinators of oilseed rape, which is the single most productive honeybee foraging crop in Britain. Oilseed rape growers and beekeepers have mutually exploited this to ensure both good seed set and abundant honey. Oilseed rape may not always be the main pollen contributing to honey production, but traces may be commonplace because honey typically has one predominant type of pollen, collected from a main source, and many other types, collected in smaller amounts from varying sources. In a study conducted by the SCRI, a hive of honeybees, situated 800 metres from a field of transgenic oilseed rape, collected ten types of pollen. Genetic analysis revealed that 70 per cent of this pollen was from the oilseed rape crop (Ramsay *et al.*, 1999).

The promoter (CaMV 35S), which is present in the gene constructs engineered into most commercially released transgenic crops, may be generally active at a low level in pollen, thereby promoting the expression of transgenes (Wilkinson *et al.*, 1997). Researchers at the University of Leicester, England, found that a marker gene (the *gus* reporter gene) expressed its protein product in the pollen of transgenic tobacco (*Nicotiana*) and thale cress (*Arabidopsis*). The protein, expressed by this transgene in pollen, remained in honey for up to six weeks. Transgenes can be expressed within pollen in honey, while novel pollen-derived proteins could remain in honey that is commercially sold (Eady *et al.*, 1995). Pollen may be an ecologically important route by which proteins expressed by transgenes are transported in the environment, either by entering the food chain in honey or by the inhalation of airborne pollen grains. The effects of pollen are well known to hay fever sufferers, and the addition of novel proteins could exacerbate allergen-related problems.

Further concern for beekeepers surfaced in May 2000, in the form of reports concerning unpublished work by Hans-Heinrich Kaatz of the University of Jena, Germany, which suggested that genes from transgenic oilseed rape could be transferred to bacteria and fungi living in the guts of honeybees (Meikle, 2000e; Lean *et al.*, 2000). In a three-year study of honeybee foraging in Saxony, Germany, Kaatz and his colleagues erected large nets over small fields of herbicide-resistant oilseed rape (a glufosinate-resistant Aventis line), so that freely foraging bees were confined. Pollen was removed from the bees' hind legs when they returned to their hives, and fed to young bees in a laboratory. The intestines of these bees were removed after set periods and the contents spread on to growth media. The microorganisms that subsequently grew were subject to genetic analysis. They were probed for the transgene that confers herbicide-resistance in the oilseed rape on which they foraged. This gene was found in some of the bacteria and a yeast cultured from the bees' guts. The bacteria had taken up the gene from the plant's pollen and incorporated it into their own genome.

The media reports that bought this work to wide public attention were based on an interview given to the German TV network ZDF, which appeared on their news programme 'planet.e' in May 2000. Professor Kaatz tried to prevent the feature being shown, while many of his scientific colleagues criticized the programme for presenting a distorted view of the research. Kaatz has subsequently had difficulty getting the study published in a major journal. In some media stories, it was implied that honeybees had taken up a gene from the pollen and incorporated it into their own genetic make-up. This was not the case. The study also showed that no harm was done to the bees from carrying the microorganisms that had incorporated the transgene; the transfer of genes to bacteria was shown to be a rare event. Nevertheless, transgenes were transferred from plant pollen to microorganisms in the guts of honeybees. It also raised the possibility, albeit fairly remote, that transgenes, including those for antibiotic-resistance used as marker genes in transgenic plants, could be transferred to other organisms, for example to bacteria living in the guts of humans via the consumption of honey.

The unforeseen impacts on non-target species described in this chapter demonstrate how little is really known about the indirect ecological effects of transgenic crops. Further unpredicted environmental impacts are certain to come to light in the coming years.

8 Engineering solutions?

The level of ecological risk associated with the 'first generation' of transgenic crops need not have been so high. These crops could have been made safer using a number of technically feasible solutions, although this would have delayed their commercial release. Transgenes could have been expressed only at particular times or in specific tissues, for example, to reduce the exposure of non-target organisms to novel proteins. Gene flow (genetic pollution) could have been greatly reduced by integrating transgenes into plant chloroplasts rather than the nucleus, and preventing flowering or the production of viable seed. In this chapter, the ecological benefits of these engineering solutions are described, along with some potential environmental and socio-economic costs.

Promoters and tissue-specific transgene expression

Promoters are regions of DNA that control gene expression. They dictate where and when a gene is expressed. A promoter is required to ensure that every transgene inserted into a plant is switched on. A promoter derived from the Cauliflower Mosaic Virus (CaMV) has been introduced into virtually every transgenic crop that has been commercially released (up to 2001). CaMV is a virus that naturally infects cauliflowers and other brassica plants. A virus is inert outside a host. It must get its genetic material inside a host's cell in order to make new viral proteins and reproduce. The promoter's role is to ensure that viral genes within a host's genome are correctly expressed. Genetic engineers have exploited the CaMV promoter system to express foreign genes in plants. They have isolated a promoter (CaMV 35S) from the virus and attached it to transgenes within gene constructs before insertion into plant cells. The promoter ensures reliable, high-level and continuous expression of transgenes.

The continuous expression of transgenes for herbicide-resistance and tolerance of pests and diseases in a wide range of plant tissue has proved effective in agronomic terms. However, in field situations the expression of transgenes in certain plant tissues provides little additional benefit, whilst creating ecological concerns. The CaMV 35S promoter is active, for instance, in the pollen of a number of transgenic crops (Wilkinson *et al.*, 1997). This means that transgenes are expressed and novel proteins are produced in the pollen, to which non-target organisms are unnecessarily exposed (Chapter 7).

It will therefore be desirable in the future to limit transgene expression to specific plant tissues or to certain stages of plant development, so that environmental exposure of transgene products is reduced. In the case of the CaMV 35S promoter, gene silencing (antisense) techniques are being developed to eliminate transgene expression in pollen. Transgenes would still be present in all plant cells, but would be expressed only in certain cell types, such as green leaves. Limiting transgene expression will lessen the risks to non-target organisms and make ecological risk assessment easier.

Potential risks due to promoters   The CaMV 35S promoter has been incorporated into most commercially released transgenic crops, but it has properties that have not been taken into sufficient consideration during risk evaluation. The unpredictable behaviour of the DNA from CaMV and related viruses (pararetroviruses) in nature, for instance, has raised concerns about the safety of the CaMV 35S promoter. This type of virus can sometimes integrate multiple copies of its DNA into plant chromosomes (Jakowitsch *et al.*, 1999). These viral genes can be transmitted from parent to offspring plants, in which they are present but inactive. Viral infection can result from inherited and latent virus genes at any time in future generations. The insertion of a viral promoter through genetic modification has triggered infectivity in a previously latent virus (Ndowora *et al.*, 1999). Viral genes can be transmitted within plant DNA in pollen, with the result that a virus can infect a plant following hybridization. Promoters and other viral genes could therefore be transferred from transgenic crops to other plants via pollen (Lockhart *et al.*, 2000). To summarize, pararetroviruses such as CaMV can integrate into plant genomes and duplicate repeatedly, become established as latent viruses in plant genomes over several generations, be induced to form infecting viruses at any time, and be spread in pollen.

Genetic exchange between virus-resistant transgenic crops and viruses has been noted previously (Chapter 4). For example, defective CaMV were able to reacquire deleted genes from the genomes of transgenic plants to become infecting viruses (Gal *et al.*, 1992). Novel and virulent viral strains can potentially evolve as a result of virus–plant recombination. However, virus–plant recombination will not be restricted to virus-resistant crops. The viral-derived CaMV 35S promoter is ubiquitous in transgenic crops. The exchange of genetic material between CaMVs involves the promoter: it is a potential recombination hot spot. Infecting viruses could therefore interact with the promoter in many types of transgenic crop, suggesting that modified DNA regions could be rapidly spread among viruses in the environment. The CaMV 35S promoter could also potentially recombine with a wide range of viruses in other organisms, including insects and mammals, to create virulent new diseases. An outside possibility is that eating fresh transgenic fruit and vegetables that contain the CaMV 35S promoter, such as the Flavr Savr tomato, could lead to recombination between the promoter and viruses in the human gut, such as hepatitis B, to produce novel and more virulent viruses (Cummins, 1994).

It is often stated that cauliflowers infected with CaMV are perfectly safe to eat. However, the isolated CaMV 35S promoter integrated into recombinant DNA is a totally different proposition. The promoter has been assumed to be safe, based on assumptions from natural CaMV infections. Little work has been done on the health or ecological risks associated with the CaMV 35S promoter itself. In an isolated and controversial study, conducted by Arpad Pusztai, it was suggested that the CaMV 35S promoter could be damaging to the health of rats. The animals were fed on transgenic potatoes incorporating a lectin gene, conventional potatoes to which lectin had been added, and potatoes without lectin. The transgenic potatoes were found to cause more damage to rat guts and internal organs than potatoes enriched with lectins by other means. However, the data were too inconclusive to support definite conclusions regarding gene constructs containing the promoter (Ewen and Pusztai, 1999). Further experiments, for example using potatoes modified with the construct but not the lectin gene, should be carried out. A number of other possibilities that might explain the results have also been suggested. The transgenic and non-transgenic potatoes differed in other biochemical measures, for example in their starch, protein, and sugar content; while differences in glycoalkaloid levels (not measured), potentially toxic chemicals naturally found in potatoes, may have accounted for changes in gut morphology. Meanwhile, in Pusztai and Ewan's published work, a relatively small number of rats were used and these probably became malnourished under experimental conditions (Coghlan et al., 1999). Nevertheless, the study raised genuine concerns. More research is badly needed in this area. However, the multinationals developing transgenic crops have funded relatively few long-term feeding studies. Government funding is also scarce. Apart from the work done by Pusztai's team, which was at a preliminary stage, the British government has funded few such studies on the biosafety of GM food. Pusztai has been unable to complete his research programme, because pressure from the scientific establishment (via the Royal Society) and biotechnology corporations led to his enforced retirement from the publicly funded Rowett Institute in Aberdeen, Scotland. When it comes to transgenic crops, the CaMV 35S promoter is the great unknown. It is rarely an issue when these crops are reviewed for release to the environment, little work has been done on its safety, and yet the potential for health and environmental impacts is considerable.

Alternatives to CaMV promoter A range of alternatives to the CaMV 35S promoter is being developed. Promoters in the future should ensure that transgene expression is limited to the green tissues on which insect pests feed, or excluded from fruits and other edible produce so that foreign proteins do not enter the human food chain. For instance, aphids are important crop pests that feed from the plant's phloem, the tissue that transports food materials such as sugars from where they are produced (e.g. leaves) to where they are needed (e.g. growing points). A phloem-specific promoter was used successfully in laboratory trials, in conjunction with a lectin gene, to confer resistance to an aphid pest (*Sitobion avenae*) in transgenic wheat (Stoger *et al.*, 1999). Insects that do not draw nutrients from the phloem have minimal exposure to the insecticidal product of this transgene. The lectin gene was present in cells throughout the plant, but it was active only in the tissues on which aphids feed.

A number of similar genetic switches are being developed to modify transgene activity. Some of these promoters can be switched on in response to externally applied chemicals. In this way, when a simple and relatively non-toxic organic chemical is applied to a crop, the transgene is switched on. A system using ethanol, for example, provides a model for this approach. The common fungi *Aspergillus nidulans* naturally biodegrades ethanol by producing an enzyme called alcohol dehydrogenase. The gene responsible for this (the *alc* regulator) can be linked to other transgenes and incorporated into a transgenic plant. When plants are treated with ethanol at low concentrations, transgene activity is promoted (Caddick *et al.*, 1998).

Transgenic plants in the future are likely to be engineered to allow more sophisticated control of gene expression. There may be a complementary move toward manipulating existing genomes, rather than introducing foreign DNA, based on a growing knowledge of how genes are expressed. The term transgenomics has been coined for this approach. The ecological problems arising from the introduction of transgenes could therefore be reduced, although the multinational corporations developing transgenic crops might decide that such modifications are neither necessary nor cost-effective. In the meantime a number of more immediate concerns about the risks associated with the 'first generation' of transgenic crops must be addressed.

Antibiotic-resistance marker genes

The genetic transformation of plants is an inefficient process. Often, only a small proportion of material will incorporate transgenes in a satisfactory manner. Marker genes are therefore essential when developing transgenic crops, to identify the material that has taken up foreign DNA. Selectable marker genes are used to select transformed material from untransformed material. The most commonly used selectable marker genes to date have conferred resistance to antibiotics. When plant material is placed on propagation medium containing the relevant antibiotic under tissue culture conditions, the transformed material can be selected from unmodified material because it is the only material that starts rooting and shooting. The material containing genes for antibiotic-resistance will also contain the other transgenes present in the gene construct.

Risks due to antibiotic-resistance Microorganisms naturally produce chemicals called antibiotics as a defence against invading bacteria. This has in turn selected for bacteria with antibiotic-resistance mechanisms. Antibiotic-resistance genes, located on bacterial plasmids, are readily exchanged between bacteria, and rapidly spread throughout bacterial populations (Talbot *et al.*, 1980). Genes for antibiotic-resistance have been routinely inserted into the gene constructs used in transgenic crops. The genes used confer resistance to a range of antibiotics, including kanamycin, neomycin, and ampicillin. These antibiotics are in clinical and veterinary use in many countries. Antibiotic-resistance genes present no direct risks to humans or animals, and the indirect risks may be minimal (Bryant and Leather, 1992; Nap *et al.*, 1992). However, there has been growing concern that antibiotic-resistance genes could be transferred to bacteria living in the guts of humans or animals, which could reduce the efficiency of antibiotic drug treatments.

Most of the DNA entering the guts of animals and humans is broken down, the acidic conditions present acting as a natural barrier. However, animal feeding studies have shown that small amounts of DNA can remain active in the gut for short periods, giving adequate opportunity for bacteria to incorporate DNA fragments into their genetic material (Webb and Davies, 1994). For example, when viral DNA was fed orally to mice, some of it persisted and was recovered from the faeces (Schubbert *et al.*, 1994). Foreign genes present in GM food – including antibiotic-resistance ones –

could theoretically transfer to gut bacteria. Dutch researchers have used an artificial human gut to monitor DNA persistence, and were surprised to find that DNA remained intact for several minutes in conditions mimicking the large intestine; for example, DNA from bacteria engineered with antibiotic-resistance genes had a half-life of six minutes (MacKenzie, 1999). DNA given to mice and rats in feeding studies has also been detected in the blood system, where it passed into organs such as the spleen and liver via the intestinal wall, and into mice foetuses via the maternal blood (Schubbert *et al.*, 1997; Schubbert *et al.*, 1998). It has also been suggested that antibiotics might even protect DNA from degradation in the gut (Schubbert *et al.*, 1994). Selection pressure during antibiotic treatments will favour bacteria that have acquired resistance genes, causing them to become predominant in the gut.

There has been a worrying increase in levels of antibiotic-resistance in disease-causing (pathogenic) bacteria, brought about by the over-use of antibiotics in human medicine and in animal feed. Antibiotics such as kanamycin have been routinely given to animals, to keep them in good health and to improve feeding efficiency, so they need less food to reach marketable size. The same antibiotics are often used to treat humans and animals. In the USA, for example, penicillin and chlorotetracycline are used as growth promoters in animal feed, even though they are also used in human medicine. Nearly 70 per cent of antibiotics produced in the USA are fed to animals as growth promoters. Antibiotic-resistant bacteria originating from pig farms have been found in groundwater. If antibiotic-resistance genes get into bacterial populations in nature there are few limits to how far they could spread. There is a growing body of evidence to suggest that antibiotic-resistance genes can spread to bacteria living in the guts of humans (Bonner, 1997; Chee-Sanford *et al.*, 2001).

Antibiotic-resistance genes from transgenic crop sources can only exacerbate the situation. The use of unprocessed transgenic crops in animal feed has prompted particular concern; for example, *Bt* maize (corn) containing kanamycin-resistance genes is widely used in livestock feed. This could lead to kanamycin and related antibiotics becoming less effective at treating disease in cattle. Kanamycin is also one of the last-resort drugs for treating multi-resistant tuberculosis in humans, while ampicillin is used in combined antibiotic therapies. The Advisory Committee on Novel Foods and Processes (ACNFP) in the UK recommended that the safety evaluation of antibiotic-resistance marker genes should include an assessment of the clinical use of the antibiotic, the likelihood of transfer of resistance genes into, and expression in, gut microorganisms, and the toxicity of gene products (ACNFP, 1994). The presence of antibiotic-resistance genes has been one of the main reasons why transgenic crops have stalled at the marketing approval stage in Europe. The European Union (EU) passed regulations in February 2001 requiring that antibiotic-resistance marker genes be dropped from commercial products by 2004 in Europe.

Alternatives to antibiotic-resistance   It is therefore desirable for antibiotic-resistance marker genes to be either excised from genetically modified organisms (GMOs) or be phased out and replaced by alternative marker systems. They could be removed from GMOs, soon after their role as selectable markers has been fulfilled, using enzymatic gene excision systems. One gene excision method, called site-specific recombination, involves the addition of short DNA regions with specific coding sequences to both sides of a marker gene. A recombinase enzyme, expressed by another inserted transgene, recognizes the specific sequences of the flanking DNA and when activated cuts out the marker gene. A number of systems of this type are being development (Ow, 2001).

A number of alternatives to antibiotic-resistance marker genes are also available, or may soon be. These include reporter genes, genes that confer resistance to cytotoxic agents, and genes that confer an ability to utilize compounds that are normally inaccessible. These alternatives may be accompanied by their own environmental risks. Meanwhile, intellectual property rights could determine how they are used (DETR, 2000).

Reporter genes are a type of marker gene that cause genetically modified material to produce a detectable colour change or become fluorescent. These genes were widely used in the early days of genetic engineering, but were replaced by selectable marker genes. Reporter genes may come back into fashion to a limited extent now that concerns have been raised about antibiotic-resistance genes. The safety of reporter genes is well characterized and they are not subject to patent protection. Reporter genes code for enzymes that bring about visible changes in the transformed material when an appropriate chemical (substrate) is supplied. The changes may be directly visible to the naked eye or visible under ultraviolet light. Reporter genes include ones that express the enzymes ß-glucuronidase (*gus*), chloramphenicol (*cat*), and luciferase (*lux*). Genes for green fluorescent protein (gfp), originally isolated from fireflies and jellyfish, make GMOs, from bacteria, to plants, to fish and rabbits glow in verdant hues. A jellyfish gfp gene was even incorporated into ANDi, a rhesus monkey and the world's first transgenic primate (born 2 October 2000 at the Oregon Regional Primate Research Center, USA), although the gene failed to function correctly.

Auxotrophic markers enable plant material to utilize an otherwise in-accessible nutrient. The sugar molecule mannose, for instance, cannot normally be utilized by plant cells. A gene that expresses the enzyme phos-phomannose isomerase (PMI) can therefore be included as a selectable marker. PMI catalyses the conversion of mannose 6-phosphate to fructose 6-phosphate, enabling plant cells to utilize mannose. They can then be selected from untransformed material when cultured on mannose-rich tissue culture media (DETR, 2000). Syngenta have patented this new marker gene system, under the name Positech. In the commercial world it is seen as the main alternative to antibiotic-resistance as a marker gene system (RAFI, 2001a).

Herbicide-resistance transgenes can be used as selectable markers in addition to being used for weed management purposes. This has most relevance to herbicide-resistant crops, because the genes of interest could be used for both purposes. In other crops, the presence of herbicide-resistance genes may be less desirable, given that such genes are known to stack up owing to transgene spread in the field. The presence of herbicide-resistance selectable marker genes might also lead to growers using the trait inappropriately. Novartis *Bt* maize, designed to be resistant to the European corn borer, was engineered with a gene for glyphosate-resistance as a selectable marker. However, it could also be used to protect the crop against this herbicide, although the company did not initially promote the transgenic crop as being herbicide-resistant.

Monsanto (Pharmacia) was awarded a patent in January 2001 covering techniques involving the insertion of antibiotic-resistance marker genes into transgenic organisms in the USA. They may now claim royalties on this commonly used genetic engineering tool. Considered together with the Syngenta patent, this means that a technique essential to genetic engineering is now controlled to a considerable extent by two of the main corporate players. The development of a range of alternatives to antibiotic-resistance is therefore desirable, including methods that are not covered by patents, to allow unrestricted research in publicly funded plant breeding institutes (RAFI, 2001a).

The manipulation of flowering

The remainder of this chapter will be concerned mainly with methods for preventing plants passing on their genes to future generations. With the mounting evidence of gene flow (genetic pollution) from transgenic crops into the environment, the development of plants that do not spread trans-genes in inappropriate ways has become of increasing importance.

Flowering could be manipulated to reduce unwanted gene flow. The molecular basis of the flowering process has been worked out in recent years. This knowledge could be applied to produce transgenic crops that do not flower, or have altered flowering times, or have modified flower structure. Flowering could be prevented altogether, with a supplementary mechanism to switch it back on for the production of seed by plant breeders. This would have limited application, however, as the production of pollen is necessary for crops that produce seed or fruit. A change in flowering time engineered into a crop could make it flower before or after other crops or wild relatives, thereby reducing cross-fertilization. Flower structure could be altered to make crops unattractive to pollinating insects, for example by modifying colour or scent. This would reduce the exposure of non-target insects to transgene products. Cleistogamy is the production of flowers that develop normally, but do not open, resulting in self-fertilization. Pollen is unlikely to escape from these flowers to fertilize other individuals, and cleistogamy is being engineered into some transgenic plants (DETR, 2000).

Male sterility could be engineered into a crop so that its flowers have infertile anthers. Pollen development can be prevented by destroying part of the anther using a gene for an enzyme called non-specific nuclease, linked to a cell-specific promoter. A backcross can restore male sterility for plant breeding purposes. External applications of chemicals might also be able to switch male sterility back on when required. Male-sterile plants can still be fertilized by incoming pollen to produce hybrid plants, however, so this represents only a one-way barrier to genetic exchange (DETR, 2000).

In maize, a recent advance has been made through the study of teosinte, a wild cousin that has a natural ability to block foreign pollen. Jerry Kermicle and his colleagues at the University of Wisconsin noticed that teosinte growing in maize fields did not pick up genes from the maize. Conventional breeding methods were used to incorporate genes from teosinte into maize. These new lines resisted pollen containing transgenes from nearby transgenic maize. This maize seed could be sold to organic farmers who want to avoid genetic contamination of their crops. Transgenic maize grown for animal feed or industrial uses, similarly, would not contaminate this new maize, even if it were growing nearby (Boyce, 2000b).

The manipulation of flowering is a pre-fertilization barrier to cross-fertilization. Post-fertilization barriers involve mechanisms that prevent viable seeds being produced.

Hybrid seed

Seed companies want to protect their new crop varieties, which represent a large investment in both time and money. They understandably want to make a satisfactory return on this investment, without others taking unfair advantage of their efforts. Crop seed has for a number of years been covered by a system of Plant Breeders' Rights (PBRs). This system gives plant breeders exclusive rights over a period of time, during which they are protected against competition and unauthorized use of their seeds. PBRs do not prevent farmers from replanting seed obtained from crops they have grown. The replanting of seed has long been acknowledged as a basic Farmers' Right, recognizing the contribution of farmers in maintaining and improving crop varieties over many generations (Juma, 1989).

However, seed companies have for some time sought ways of preventing farmers from recycling seed from new crop releases. The first step to achieving this occurred in 1908 when George Schull discovered the process of hybridization. Hybrid seeds are the first generation (F_1) offspring of two distinct parental lines. Seeds of hybrid crops do not breed true, and if planted will give only low yields. Farmers wanting to maintain high yields need to purchase new seed annually. Plant protection is therefore embedded in the seed. Commercial hybrid maize was first marketed in 1924 by Henry A. Wallace, who soon after formed Pioneer Hi-Bred, which is still one of the world's largest seed companies. Therefore, hybrid maize was specifically invented to create a copy-protected product. Any yield increases that may have occurred were incidental to this main aim (Steinbrecher and Mooney, 1998). Although marketed through the concept of 'hybrid vigour', as a high-yielding technological breakthrough, it has become apparent that conventional breeding might have produced maize cultivars with similar or even better performance, if given the same research investment. Today, virtually all the maize seed sold worldwide is hybrid, and all this seed is controlled by a handful of giant multinational corporations (Lewontin, 1993).

Not all crops have been as amenable as maize to becoming hybrid. Despite massive research efforts during the 1980s, small-grain cereals such as rice, wheat, barley and oats, and leguminous crops such as soyabean, resisted attempts to make them hybrid, although there have been some recent successes. Millions of dollars have been invested in produced hybrid wheat, for example, and Monsanto (Pharmacia) and Syngenta may soon have commercially viable hybrid wheat seed. Wheat is grown on a larger land area than any other crop, so the financial gains from hybrid wheat could be enormous (Steinbrecher and Mooney, 1998).

Terminator technology

Since the early 1980s, seed companies have increasingly regarded the saving and replanting of seed as theft of their intellectual property. This attitude has recently hardened with changes in patent laws, which give full patent protection to transgenic crop seed. Because many transgenic crops, such as Roundup Ready soyabeans, are not hybrid, the companies holding the patents on them have vigorously policed their use. Gene licensing agreements signed by farmers expressly forbid them to save transgenic seed for replanting. Monsanto have the right, through the agreement, to inspect farms at will and has taken legal action against more than 100 soyabean growers whom it considers to have violated licensing agreements (RAFI, 1997). Monsanto has taken out adverts in farming journals to explain their position and to plead with farmers to respect the licensing agreements (Anon., 1997). The system is open to abuse, however, because seeds can potentially be replanted.

In early 1998, a new genetic engineering technique was announced that provided an elegant system of built-in patent protection and seemed to offer the perfect solution for the seed companies. The technology rendered a plant's seeds sterile. This was seen as an extension of the hybrid seed approach. Hybrid seed can be bred from, even if it does not breed true, but sterile seeds have an advantage in that genetic material cannot be retrieved from the crop. Sterile seeds would also have ecological benefits, by preventing unwanted gene flow from transgenic crops via their seeds. However, sterile seed technology was not developed with ecological aims uppermost, but as a means of protecting the patent rights of multinational companies. The Rural Advancement Foundation International (RAFI), a Canadian-based non-governmental organization, was quick to see how the technology could have serious implications for farmers, particularly in developing countries. They called it 'Terminator technology', and the name stuck (RAFI, 1998).

How Terminator works   The United States Department of Agriculture (USDA) and a Mississippi-based seed firm, Delta and Pine Land Company, were awarded a patent in March 1998 that covered a broad range of applications of a technique for sterilizing crop seed using genetic modification. The technique involved the introduction of a toxin gene that is present in all the plant's cells, but which becomes activated only in the seeds. This is achieved using a promoter called *Late Embryogenesis Abundant* (LEA), which turns on the toxin gene during the later stages of seed development. The LEA promoter and toxin gene are integrated into a plant within the same gene construct. In the patent, the favoured toxin gene is one that encodes for ribosome inhibitor protein (RIP), obtained in this case from the soapwort plant (*Saponaria officinalis*). The toxin acts by interfering with the cell's protein synthesis machinery, preventing the production of any protein in developing seeds (Crouch, 1998).

Crop plants containing the LEA and toxin gene construct will therefore grow normally, with seeds that look indistinguishable from normal seeds. However, if planted these seeds will not germinate. This system is well and good, but plant breeders need to grow many generations of a transgenic crop to obtain commercial quantities of seed. These plants will need to contain the seed-specific toxin gene construct, but in a completely inactivated form. The solution is provided in the patent, through additional layers of control. A piece of blocking DNA is inserted between the LEA promoter and the toxin gene in the gene construct. This block prevents the toxin gene becoming activated by the LEA sequence, until it is removed using an enzyme called a recombinase. Recombinase acts to cut DNA precisely, at both ends of the blocking DNA sequence, so that it floats away, allowing the LEA promoter and toxin gene to fuse together. A recombinase gene is therefore added to the construct, but attached to a promoter sequence of its own that is permanently switched off, thus depressing the recombinase gene. If nothing is done to alter this off-switch, the blocking DNA remains in place and generations of viable seed can be harvested from the transgenic crop for plant breeding purposes. When the seed is about to be sold, however, an external chemical treatment is applied, which turns the recombinase on, thus removing the blocking DNA, and priming the toxin gene to be switched on during subsequent seed development. In the patent, the antibiotic tetracycline is suggested as one possible chemical treatment. Seeds would be soaked in chemical solutions, dried, and then packaged for sale. The removal of the blocking DNA is a one-off event and no viable seed can be obtained in the future from treated plants or their progeny (Crouch, 1998).

Environmental concerns Sterile seed technology may prove ecologically useful in preventing unwanted gene flow via the seeds, but pollen containing toxin genes could still spread from these crops. Pollen carrying toxin genes could potentially infect adjacent non-terminator crops, and cause infertile seeds to be produced. Every pollen grain released from a terminator plant will contain the LEA promoter and toxin gene construct. If this pollen fertilizes another plant, any offspring will also contain these transgenes. The seeds of these plants will look normal, but may not germinate. Farmers may be unaware of any genetic contamination in their crops (Crouch, 1998).

The external chemical switch, applied just before the seed is sold, is unlikely to be 100 per cent effective. This could result in a small proportion of plants making viable seeds. These could remain in the soil to produce 'volunteer plants' in the following year's crop, which could pass transgenes to other crops or wild relatives. Genes within plants are also naturally switched on and off through gene-silencing mechanisms. Terminator genes could become activated by such a mechanism in future generations, destroying the seed viability of the plants they happen to be in (Crouch, 1998).

Socio-economic implications The main concerns expressed about Terminator technology, however, are to do with food security. By threatening to eliminate the age-old right of farmers to save seed from their harvest, the technology may jeopardise the food security of 1.4 billion people. Half the world's farmers are poor and cannot afford to purchase seed every growing season, yet they grow 15–20 per cent of the world's food, feeding over one billion people. Seed collection, replanting, and exchange are vital for these poorer farmers. If Terminator technology becomes widely utilized in the 78 countries in which patents have been applied for it, multinational seed and agrochemical industry will gain an unprecedented capacity to control the world's food supply (RAFI, 1998).

The saving of crop seed is commonest in the developing world, but it also occurs on farms in the developed world. Some estimates have suggested that, traditionally, over 20 per cent of all soyabean fields in the US mid-west, and up to 50 per cent of soyabeans in the South, are planted with farmer-saved seed. Most North American wheat farmers typically rely on farm-saved seeds, buying new seed once every four or five years. Seed-saving in crops such as lentils and peas also occurs, with fresh seed being bought every few years (RAFI, 1998). Seed companies therefore stand to make significantly increased profits by preventing this type of replanting. Multinational corporations now control most of the companies supplying seed for the major agricultural crops, and could easily start replacing older seed varieties with transgenic seed containing Terminator technology.

In November 1998, the Consultative Group on International Agricultural Research (CGIAR) agreed that plant breeders in their 16 affiliated public research centres would not use any genetic system designed to prevent germination. These research centres include IRRI in the Philippines, CIMMYT in Mexico, IITA in Nigeria, CIP in Peru and

ICRISAT in India. 'Let's be absolutely clear,' said RAFI's Edward Hammond,

> This is a technology that deliberately sterilises farmers' fields, that offers zero agronomic benefit, that is openly aimed at the South, and that is now in the hands of a giant, aggressive multinational company with more than enough resources to follow through on the plan. If Terminator is widely deployed, we will be facing a crisis for small farmers and *in situ* conservation. (RAFI, 1998)

In a worst-case scenario, which until recently would have been dismissed as science fiction, a chemical or a self-replicating virus might be developed that disables all seeds not carrying a Terminator-type modification. The US military have shown interest in this technology, which can theoretically be used to switch any introduced trait on or off with a given external trigger. Exported seed could have a self-destruct switch that is triggered by chemical 'bombing'. This could be used against illegally grown crops, or even as a form of biological warfare (Brittenden, 1998; Steinbrecher and Mooney, 1998).

Multinational corporations and patents   The research that led to the development of Terminator technology was done at the USDA's research station in Lubbock, Texas. Melvin Oliver, one of the team who developed it, has stated that the reason for the USDA's involvement was, 'to protect American technology and make us competitive in the face of foreign competition'. The USDA has also insisted that Terminator technology would benefit poorer farmers by allowing companies to protect their interests while marketing a wider range of transgenic seed to suit growing conditions in less developed countries (Edwards, 1998; Vidal, 1998b). Terminator technology has become a sensitive issue, however, and public-sector scientists have been discouraged from working on it by the US government. It has become a technology that is exclusively in the hands of giant multinational corporations.

Delta and Pine Land took up the option to license the patented technology that was jointly developed with the USDA. Monsanto bought Delta and Pine Land in May 1998, thereby gaining control of the patent (RAFI, 1998; Freiburg, 1998). Reaction against Terminator technology led Monsanto to put it on the back burner, but the approach has continued to be developed and patented by major corporate players. Syngenta, the world's largest agribusiness as a result of a merger between AstraZeneca and Novartis in November 2000, was only days old when it was awarded a major new Terminator-type patent (it had inherited five others). The new patent covered a complex system for the chemical control of a plant's fertility. A number of recent patents have also involved chemically dependent plants ('Traitor technology'), which develop properly only after the application of proprietary pesticides or fertilizers containing chemicals that act as genetic switches (RAFI, 2001b).

Modifications involving genetic switches that can be triggered by externally applied chemicals could soon transform many areas of agriculture. International Coffee Technologies Inc., for example, are developing transgenic coffee plants that can be ripened simultaneously when sprayed with ethylene. Coffee usually ripens at different times and therefore has to be hand picked. Smallholder farmers produce around 70 per cent of the world's coffee. Ripening-controlled coffee is being developed for large plantations and will be highly profitable for big business interests. However, the use of machine harvesting in transgenic coffee is likely to drive smallholders out of business and increase poverty-related problems in the developing world, by taking away the livelihoods of millions of people (ActionAid, 2001).

Apomixis

Apomixis is a form of asexual reproduction, in which crops effectively clone themselves by producing seeds without fertilization. Many wild plants produce genetically identical offspring by this method, and several genes from such plants have been isolated and experimentally introduced into transgenic crops. When bred into maize, apomixis leads to the creation of clones. This makes the production of hybrid seed quicker and easier. This maize could open up new markets in the developing world to seed companies. Third World farmers have not been a viable market for hybrid maize, since they cannot generally afford to buy fresh hybrid seed every year. Farmers can replant seed from apomictic maize and genetically identical plants will result. In January 1997, the USDA was awarded a patent to cover techniques for producing apomictic maize.

Apomixis could help to reduce gene flow from transgenic crops. Apomictic plants can be made male-sterile, as they do not need to produce pollen in order to produce seed or fruit. In apomictic crops, the egg retains its full set of chromosomes, instead of being halved prior to fertilization, and so does not need to receive pollen from another plant. The offspring are clones of the female parent and so apomictic plants neither accept pollen from other plants nor need produce any pollen themselves (Coghlan, 2000b).

It might eventually be possible to introduce genes conferring the apomictic trait to plants without using genetic engineering. University research groups have expressed a desire that apomixis technology be made available to subsistence farmers free of charge. At a meeting in Bellagio, Italy, in 1998, for example, the pioneers of this research signed a declaration aimed at preventing multinational companies obstructing the free development of apomixis. However, the multinationals are starting to move in and have filed many patents in this area. As with all the novel technologies discussed in this chapter, intellectual property rights will determine how they are employed and who will have access to them.

Seed from apomictic crops could be bred to suit local conditions, and be saved and grown from year to year in the developing world. In this respect, apomictic crops are generally considered to be more acceptable than transgenic crops containing Terminator technology. In the long term, however, the cultivation of apomictic lines could be problematic. The conversion of a cross-fertilizing species into an apomictic cultivar means that it can no longer adapt to environmental change. They might, for instance, become more prone to disease over time than conventional cultivars. Apomictic crops stand in contrast to the locally-adapted crops (landraces) of traditional agricultural systems. These are adapted to local conditions, have high genetic diversity, and are best suited to change in environmental conditions. Apomixis therefore potentially represents an erosion of diversity and a threat to genetic resources.

Chloroplast transformation

All of the commercially released genetically engineered crops to date have had transgenes inserted into the plant's nuclear DNA. In addition to the nucleus, however, two other structures in the cells of plants have their own DNA: chloroplasts and mitochondria. Chloroplasts are responsible for photosynthesis, a process in which energy from sunlight is used to convert atmospheric carbon dioxide into the sugars that the plant uses for growth. Mitochondria are responsible for energy production in the cell. It is thought that chloroplasts and mitochondria were at one time free-living microorganisms that entered into close symbiotic relationships with the single cell ancestors of today's plants (Margulis, 1993). The genetic material of both chloroplasts and mitochondria remains self-contained and, unlike nuclear DNA, no genetic recombination occurs during reproduction. In higher plants, inheritance of chloroplast characteristics is usually maternal, with offspring inheriting their chloroplasts from the female parent only.

A promising option to reduce inappropriate gene flow is therefore to insert transgenes into the DNA of chloroplasts rather than that of the nucleus. It is much more difficult to do this technically and may not be suited to all transgenes. However, once it has been achieved, transgenes will be much less likely to be transferred in pollen because chloroplasts are inherited maternally, passing down the generations in the egg cells, and are not present in the male pollen cells of most species of higher plants. Another advantage of chloroplast transformation is that it is possible to be more precise about where the transgene is inserted, compared to current transgenic transformation techniques. There is less DNA in a chloroplast than in the nucleus, while the complete gene sequences of chloroplasts for many crop species have already been mapped, ahead of nuclear genome sequencing. One nucleus occurs in a cell, compared to many thousand chloroplasts, and the high copy number of transgenes in chloroplasts enables high levels of expression to be achieved (Daniell *et al.*, 1998; Daniell and Varma, 1998).

One of the most thoroughly studied systems for the inheritance of herbicide-resistance involves a mutant chloroplast gene (*psbA*), which confers resistance to triazine herbicide (Souzha-Machado *et al.*, 1978). Crossing herbicide-resistant plants with related weeds did not alter the maternal inheritance of herbicide-resistance (Daniell and Varma, 1998; Ali *et al.*, 1986; Darmency, 1994). It was considered unlikely that the chloroplast gene would be transferred to weedy relatives in pollen. The assumption is that transgenes will be similarly stable and inherited only maternally if integrated into chloroplasts (Daniell *et al.*, 1998).

Herbicide-resistance and insect-resistance traits have been engineered into plants using chloroplast transformation (Ye *et al.*, 1990; Daniell, 1997; Daniell, 1999). Glyphosate-resistance, for example, has been achieved in tobacco. The trait was not transferred via pollen, but was passed on maternally to offspring via the seeds (Daniell *et al.*, 1998).

High expression levels of toxins have been reported when *Bt* genes have been integrated into chloroplasts. In one study, a *Bt* toxin gene (*Cry1A*) expressed in the chloroplasts of tobacco caused 100 per cent pest insect mortality (Kota *et al.*, 1999). In another study, leaves of experimental tobacco plants, with up to 10,000 coding sequences of the *Bt* toxin gene in each chloroplast, accumulated an unprecedented 3–5 per cent soluble *Bt* insecticidal protein. These plants were extremely toxic to lepidopteran larvae, specifically tobacco budworm (*Heliothis virescens*), corn earworm (*Helicoverpa zea*) and an armyworm species (*Spodoptera exigua*) (McBride *et al.*, 1995). Chloroplast transformations that produce plants with high levels of toxin in their leaves could prove useful in managing the development of insect resistance to *Bt* toxins. Using chloroplast transformation enables the *Bt* toxin gene to be expressed predominantly in the green tissue of a plant, and not in the pollen. This may be an important consideration in the light of results discussed elsewhere showing that *Bt* in pollen causes mortality of non-target insects, including monarch butterflies (Chapter 7).

In oilseed rape (*Brassica napus*), a number of studies have demonstrated the transfer of transgenes via pollen from transgenic crops to wild relatives (Chapter 6). A study of the movement of genes located on chloroplasts from oilseed rape to 18 distinct populations of wild turnip, along a section of the River Thames in England, showed that there was very little pollen-mediated exchange of chloroplast genes over a three-year period. Susan Scott and Mike Wilkinson, of the University of Reading, concluded that, from known crop–weed hybridization rates, gene flow would be rare if plants were genetically engineered via the chloroplasts (Scott and Wilkinson, 1999).

The use of chloroplast transformation could therefore considerably reduce gene flow, and it is likely to become an increasingly used technique (Gray and Raybould, 1998). However, there are limitations to its use. Chloroplast transformation does not stop pollen coming from weedy relatives outside a crop to fertilize plants, and this route could still form hybrid plants. In addition, some plant species can have chloroplasts in their pollen, so this is not a universal solution to preventing gene flow.

Critics of genetic engineering dismiss many of the methods discussed in this chapter as 'technological fixes'. Sue Mayer of GeneWatch has remarked that, 'It's extraordinary the ends people are expected to go to accommodate GM'. She was commenting in particular on the maize hybrid barrier method, in which the onus was on the organic growers to buy special seed that would resist transgenic pollen (Boyce, 2000b). Similarly, oilseed rape growers in Canada have been told that it is up to them to monitor what transgenic crops are being grown in the area, so as to minimize potentially damaging transgene spread. The real solution in cases involving organic and seed crop production is not provided by 'technological fixes'; it is to stop the growing of transgenic crops in these areas.

As one 'fix' is added to another, additional complexity is added that may increase the unpredictability of genetic transformation in the field. It should probably be accepted that certain crops present too much risk to be considered for commercial release, and that 'technological fixes' are not the answer. Moreover, many of the major concerns about genetically engineered crops are not addressed by engineering solutions. In some cases, the socio-economic, ethical and political problems associated with genetic engineering may be exacerbated by these technological 'solutions'. This is the case for Terminator technology.

If it is accepted that genetically engineered crops do have a role to play in agriculture, they could be made safer by the application of techniques such as some of those described in this chapter. This will ensure that genetically engineered crops in the future present fewer ecological risks than the 'first generation' of transgenic crops. Even if ecological risks are reduced, however, transgenic crops may still be costly, in terms of threats to crop diversity, ecological biodiversity, and a range of undesirable socio-economic impacts.

9 Trees

Forests cover around one third of terrestrial landscapes. They provide an invaluable source of fuel, building material for shelter, wood pulp for paper and packaging, and some food. Most forest products have until recently been harvested in a relatively sustainable manner. However, the per capita increase in consumption of forestry products rose by around 50 per cent in developed countries, and 300 per cent in developing countries, between 1970 and 1994, and continues to rise (FAO, 1997a). Transgenic trees are seen as one means of increasing the productivity of managed forests. This chapter outlines, on the one hand, how transgenic trees could provide economic and environmental benefits, and, on the other, why there is considerable potential for ecological risk.

Increased growth rates

Tree breeding programmes have traditionally focused on obtaining fast, straight growth with minimal branching. The long generation time of trees has been a major constraint on selective breeding, although the situation for some timber trees has improved in recent years. The rotation time for *Eucalyptus* (gum trees), for example, has been reduced from fifteen to seven years. Genetic engineers expect to decrease generation time further. Transgenic trees are being designed for intensively managed plantations, where quicker growth could increase production through earlier harvesting. It has been claimed that this could help to meet rising demand for timber and wood pulp products, while taking the pressure off native forests, thereby helping to conserve biodiversity.

In April 1999, Monsanto (Pharmacia) teamed up with two American companies, International Paper and Westvaco Corporation, and the New Zealand-based Fletcher Challenge Forests, to form a joint forest biotechnology venture called ArborGen. Their stated aim was to produce and market transgenic tree seedlings. These companies linked up with Genesis Research and Development, New Zealand, to acquire intellectual property rights covering forestry biotechnology. ArborGen initially focused on raising the productivity of common wood pulp trees, such as *Eucalyptus*, poplar (*Populus*), loblolly pine (*Pinus taeda*), Monterey or radiata pine (*Pinus radiata*) and sweetgum (*Liquidambar styraciflua*).

Monsanto had previously established a joint venture with ForBio, an Australian plant biotechnology company, to establish automated mass propagation techniques and to reduce the rotation time for species such as teak (*Tectona grandis*), acacia (*Acacia mangium*), and *Eucalyptus*. Another joint venture, between Fundación Chile, Interlink Associates (USA) and Silvagen (Canada) aimed to develop transgenic Monterey pine for faster growth rates, better wood pulp quality, and enhanced pest and disease-resistance. Field trials in Chile to assess resistance of transgenic trees to moth pests commenced in 2000. Further joint ventures, aimed at exploiting the possibilities of transgenic trees, are certain to be announced in the coming years.

Tree sterility The most promising method for increasing the growth rate of trees is to introduce sterility. Reproduction is a costly process. A tree typically uses 15–30 per cent of its total energy on reproductive structures, such as flowers, cones and fruits. By preventing a tree from reproducing, all its energy can go towards growth. Transgenic poplars in field trials, for example, have grown up to four times faster than the conventional trees used to produce wood pulp for newsprint.

Transgenic tree plantations will enable more wood to be grown on less land, but the environmental benefits may be accompanied by significant ecological costs. Good management practices in plantations can produce sustainable yields, maintained over time without affecting the soil's nutrient status, structure, or hydrological function. But fast-growing transgenic trees will make additional demands on soil nutrients and water, with consequences for the long-term fertility of soils. Substantial fertilizer inputs might be necessary to maintain high yields.

Most of the field trials to evaluate enhanced growth and sterility have been conducted in Asia and Oceania. The first commercial plantings of such transgenic trees probably began in Indonesia, on Kalimantan and Sumatra, in 1999. Little attention has been paid to the potential environmental impact of these transgenic trees (Owusu, 1999).

Low-lignin trees

Poplars (cottonwoods and aspens) have become favoured forestry trees for producing wood pulp, because of their rapid growth rates. However, the lignin content of poplars creates yellow paper, rather than the desired white. Treatment with chlorine is therefore necessary to remove the lignin from the wood fibres and to bleach the paper. It is an expensive process and chlorine is environmentally damaging. Sawmills in North America and Europe are required to spend large amounts of money cleaning up after the chemical treatment process.

Lignin is a naturally produced woody substance that makes plant cells rigid, and provides tree with the strength to stand upright. Lignin content may soon be reduced by around 20–30 per cent in transgenic conifers, and 15–20 per cent in transgenic broadleaf trees. This could reduce the amount of chemicals used in pulping and bleaching by around 30 per cent, making processing cheaper and easier. Therefore, fast-growing low-lignin transgenic trees could significantly enhance both the quantity and quality of wood pulp, while using less energy and producing less toxic effluent.

The ecological costs of low-lignin trees need to be considered, however, alongside their environmental benefits. By introducing genes that reduce the amount of lignin a tree produces, its ability to stand firm will be compromised. This is less of a problem in intensive plantations, because trees are grown at high densities, minimizing wind exposure, and are harvested young. It would be more of a problem for trees grown in less dense plantations. Zeneca (now part of Syngenta) produced transgenic low-lignin poplars through a process of trial and error, for example, to strike a balance between whiter wood pulp and retaining the tree's ability to withstand the English weather. These poplars were released in Britain's first transgenic tree field trial in July 1999, but they were broken by activists rather than the wind (Brooks and Brown, 1999). Zeneca's poplars were female and so could not have produced pollen. However, if transgenes responsible for reducing lignin content were to get into native tree populations from commercial plantations, then hybrid trees could be left vulnerable to damage by strong winds.

Lignin also forms part of a tree's natural defence system, by reducing its digestibility to herbivores. The benefits of low lignin therefore have to be balanced against increased susceptibility to pest attack. Additional pesticides may have to be sprayed, at economic and environmental cost.

Herbicide and pest resistance

The fast-growing transgenic trees produced by the ArborGen and ForBio joint ventures will incorporate genes for resistance to Monsanto's Roundup® (glyphosate) herbicide. This will provide benefits for weed control when tree seedlings are becoming established in nurseries or plantations. If increased amounts of herbicide are sprayed on transgenic trees, however, it could be detrimental to the environment. All vegetation other than tree seedlings can be killed in nursery stands, with adverse effects on biodiversity. Aerial spraying is often carried out on trees, so even herbicides with minimal drift in field crop situations can fall on land around forests, and potentially pollute waterways. Meanwhile, regular use of one herbicide, in this case glyphosate on seedling stands, could select for herbicide-resistant weeds, with the result that the benefits gained may decline over time (Owusu, 1999).

Transgenic trees modified for insect-resistance, with genes from *Bacillus thuringiensis* (*Bt*), have been grown in field trials. The tree species involved have included poplar, spruce (*Picea mariana*), walnut (*Juglans nigra*) and apple (*Malus domestica*). As with transgenic crops, resistance management strategies will be necessary if insect-resistant trees are to be effective in the long term, although the longer generation time of trees makes this of less immediate concern. The development of trees resistant to a range of other pests and diseases is also ongoing; for example, a group at the University of Dundee in Scotland are seeking to make elms (*Ulmus* spp.) resistant to Dutch elm disease.

Many experimental trials with transgenic trees have been conducted in the USA, mainly in Oregon, California, and Washington states. Large-scale plantings of poplars account for around half of these trials, which have involved herbicide-resistant and pest-resistant trees. In South America, trials have been done chiefly in Chile and Uruguay, with *Eucalyptus globulus*, *E. grandis*, and *Pinus radiata*. The two main traits examined have been herbicide-resistance and lignin modification. By 2000, experimental field trials with transgenic trees had also been conducted in Brazil, Canada, China, Australia, New Zealand, Indonesia, Japan and several countries within Europe.

Orchard trees

Although most genetic modifications to trees have been done with the aim of improving timber and wood pulp production, over 20 per cent of experimental trials have involved orchard species. The genetic modification of fruit trees, including apple, orange (*Citrus*), olive (*Olea europea*), papaya (*Carica papaya*), pear (*Pyrus communis*), peach (*Prunus persica*) and plum (*Prunus domestica*), has been done with the aim of increasing the quantity or quality of the crop.

Orange trees (*Citrus sinensis*) have been engineered to produce fruit as early as their first year, using a transgene from *Arabidopsis* that promotes flowering, instead of during year six or later. This technique could increase productivity and enable breeders to speed up the genetic improvement of citrus and other woody species (Peña *et al.*, 2001). Citrus has been experimentally modified, for example, to neutralize the bitter liminoids that can lower the quality of fruit juice. In this case, more insecticides may need to be sprayed on orchards as liminoids evolve to defend trees against attack by pests and diseases. Citrus is also being modified to be resistant to pests and virus disease, and to make the fruit easier to peel.

Orchard trees produce food, which is more likely to be rejected by consumers than timber and wood pulp products from transgenic trees. The latter are also being developed initially for commercial production outside Europe – in Latin America and South-East Asia – where there has been less opposition to genetic engineering. As far as biotechnology companies are concerned, therefore, forestry provides ideal opportunities to move genetic engineering forward.

One area of transgenic fruit production that may flourish, even if consumer opposition to GM food continues, is the development of vaccines in fruits. Transgenic bananas (*Musa*) have already been developed for this purpose to contain hepatitis B vaccine (Kiernan, 1996). Cooking destroys the vaccine, so fruits are better than many vegetables in this respect, because they are normally eaten raw.

Gene flow

When plantations of transgenic trees are grown near wild relatives, as is often the case, the likelihood of genetic exchange occurring may be high. Trees are much closer to native species than field crops, both genetically, in that they have not been selectively bred to the same extent, and in terms of their habitats. In addition, large amounts of pollen are often produced. The height of trees, and the fact that they are often grown on hillsides or at high altitudes, can help pollen dispersal by the wind. Trees therefore have the capability of spreading their genetic material over large areas. Pine pollen, for example, can travel distances of 600 kilometres (400 miles) to reach another tree (Brown, 1999). Long-distance pollen dispersal has also been recorded for temperate conifers and tropical broadleaf species (Owusu, 1999).

Unwanted gene flow has long been a problem for tree breeders. Just as vegetable breeders need to establish minimum separation distances around seed crops, tree breeders need to protect plots of trees being grown for their seed from the incoming pollen of wild relatives. The greater distances that tree pollen can travel can make the situation for tree breeders more problematic. Small plots, however, can be covered in netting to cut down pollen flow. In the reverse situation, the spread of pollen from large plantations of transgenic trees will make it inevitable that gene flow to native populations will occur.

Trees were until recently considered to be fertilized largely by near neighbours. However, DNA paternity testing of trees using genetic markers has shown that pollination from a distance is more common than previously thought. In a six-hectare stand of bur oak (*Quercus macrocarpa*) in the USA, for example, a genetic analysis of 282 acorns revealed that in 57 per cent of cases pollen must have come from outside the study stand. As oak is wind-pollinated and its density declines with distance, the researchers concluded that for such a high proportion of distant fathers to occur, the female trees must have in some way been favouring their pollen. This could arise by outbreeding mechanisms such as embryos fertilized by the pollen of closely related and neighbouring trees being selectively aborted. A number of other studies have confirmed that wind-pollinated trees regularly interbreed over long-distances – over 100 metres or more (Holmes, 1995; Dow and Ashley, 1998).

Pollen in the atmosphere   Pollen from trees will be blown downwind from plantations, in a manner similar to the pollen from transgenic crops, for up to several kilometres. However, a small proportion of pollen from crops and trees can also be carried over much longer distances in certain weather conditions. Particles, including pollen grains, can be carried vertically up into the atmosphere by convective currents (thermals) or storm systems. Convective currents occur on warm days with low wind speed, as the ground heats up and air rises. Pollen will descend when the air cools, usually in the evening. Convection cells may be in the order of 5–10 kilometres across and last for several hours. Pollen could be kept aloft by convective currents for the best part of a day and travel up to 180 kilometres before descending to earth. Storm systems can transport particles 8–12 kilometres high, where strong winds could also carry pollen over similarly long distances in a few hours (Emberlin *et al.*, 1999; Mandrioli *et al.*, 1984). One study showed that pollen travelled 300–400 kilometres in one day over the North Sea from its source in Britain; in another study, tree pollen was found on the treeless Shetland Islands of Scotland, 250–380 kilometres from the nearest source forests (Hirst *et al.*, 1967; Tyldesley, 1973). Birch (*Betula*) pollen from woods in Sweden was transported within high-level air masses, for 9–20 hours, over long distances within Scandinavia (Hjelmroos, 1991). Pollen has even been detected in the air over the middle of the Atlantic Ocean. Once pollen has been transported into the upper atmosphere, therefore, it can potentially travel for hundreds of kilometres (Emberlin *et al.*, 1999).

Long-distance dispersal should be considered with respect to pollen viability times. The pollen of many orchard trees, for example, is viable for up to two days in favourable conditions (cool, dry, with low light intensity), but for only a few hours at most in warmer, humid, bright conditions. If pollen is viable for several hours or more, then long-distance dispersal can occur, although most pollen transported by convection cells or storm systems will probably not be viable when it descends to earth. Nevertheless, long-distance pollen transport has the capability to disperse transgenes to new areas, far removed from their site of release.

Forest diversity

Hybrid trees have been shown to arise as a result of transgenic trees pollinating native trees. In one study, 3.8 per cent of the progeny of wild females growing near a cultivated poplar (*Populus*) plantation were fathered by pollen from male plantation trees. Hybrids established and grew at about the same rate as wild seedlings (DiFazio *et al.*, 1999). If transgenes for faster growth were to find themselves in hybrid species, they might become more invasive than wild-types. They would grow faster, reach reproductive age more quickly, and therefore potentially out-compete trees lacking the transgene. Growth-enhancing transgenes could therefore become established in wild populations. Faster growth rates in natural environments could have knock-on effects in terms of the flora and fauna that native trees support. In addition, native species could be indirectly affected, for example via changing shade patterns on the floor of woods or via subtle changes in soil ecology.

Trees live for many years. Any ecological risk posed by transgenic trees will therefore be present for an extended period. The longevity of trees also means that transgenes will be subject to a much wider range of environmental stresses, such as temperature extremes and pest attack, than transgenic crops that are annuals or biennials. Stress can alter the way genes are expressed. Stress-induced side-effects on transgene expression will be much harder to detect and address in trees than in crops. There is a greater risk of mutation, while the actions of transgenes are likely to become more unpredictable over longer periods (Owusu, 1999). In Germany, a field trial of transgenic aspens (*Populus tremula*) flowered in their third year, whereas flowering is normal in year eight. However, the trees had been designed not to flower at all. The mechanism to prevent cross-pollination in this case backfired, due to an inadequate knowledge of how genetic modification affects plant physiology (Bioengineering Action Network, 2000).

The remote locations of many transgenic tree plantations may mean that little monitoring for environmental impact is done. Taken together with the longevity of trees, this suggests that additional safeguards might be required within international biosafety protocols that deal specifically with trees. In particular, these should address the potential threats to forest biodiversity.

Native trees can support abundant biodiversity, and they may be keystone species in complex ecosystems. They provide habitats in their roots, trunks, branches and leaves for numerous species, including mycorrhizal fungi, insect herbivores, birds and mammals. Many of the species that form communities on native trees are highly specialized to exploit narrow niches within that particular tree species. Several important temperate woodland species have been genetically engineered in experimental trials for increased growth rates and other traits, including European sweet chestnut (*Castanea sativa*), larch (*Larix*) and silver birch (*Betula pendula*). These are species that support large and diverse communities. Modification of these trees could therefore potentially affect many other species. Rotting trees also provide unique habitats for numerous creatures. Trees modified to be low in lignin will rot quickly, to the detriment of woodland communities. Accelerated wood decay can also affect soil structure, with knock-on effects on soil ecology.

Conventional tree breeding, over the past quarter-century, has moved towards the propagation of monoclonal planting stock, from selected individuals. Therefore, the genetic diversity of forests has been declining for some time. Genetic engineering enables a wider range of genetic improvements, but these will be done in conjunction with a cloning technique called somatic embryogenesis. Large numbers of cloned seedlings can be produced from a small amount of initial plant tissue using this technique. The use of transgenic trees will therefore continue the trend towards intensive monocultures of low genetic diversity.

Native forests are being lost at an alarming rate. Around 13.7 million hectares of tropical forest vanish annually (Langelle, 2001). In recent years, plantations have replaced much native forest and woodland. Plantations, however, are poor imitations of natural forests. Wildlife, including flower, insect, bird and mammal species, can find limited habitats and food resources in plantations compared to native forests. Intensively managed plantations often comprise non-native trees, such as exotic conifers, which support few species because they do not supply the niches that native flora and fauna have evolved to exploit. In addition, undergrowth is often cleared using herbicides, a practice likely to be more common in plantations of herbicide-resistant trees, while dead and rotting trees are removed. Therefore native wildlife survives only as a 'relic' community in plantations of this type. If large blocks of land in Europe or North America were given over to single transgenic tree lines, particularly with sterility traits that enhance growth rates, there is a risk that this situation could be exacerbated. Transgenic trees with in-built sterility will not produce flowers, cones or fruits. These are structures that many native insect, bird and animal species utilize. Sterile trees would therefore be able to support only a much diminished relic community (Owusu, 1999).

Between 1980 and 1995 the plantation area in the developing world doubled and is expected to double again by 2010. The replacement of native forest by non-native plantation trees has been particularly acute in developing countries. It is in these countries where the first commercial plantations of transgenic trees have been established. In a report published by the WWF, the author Rachel Owusu suggested that transgenic trees could soon be grown in large-scale plantations in Chile, China and Indonesia (Owusu, 1999). This will occur despite little research having been done on their environmental impacts. Meanwhile, the commercial interests behind transgenic tree plantations are keen to promote an increased use of forestry products, such as paper and packaging, instead of promoting recycling or reductions in usage, which is unlikely to take the pressure off native forests.

Transgenic trees and climate change

Forests and woodland help to maintain the planet's atmosphere, with trees acting as generators of oxygen and sinks for carbon dioxide (CO_2). Changes in forestry practice, such as the planting of intensive plantations of fast-growing trees, may therefore alter carbon cycling at the biosphere level. This could play a limited role in moderating the effects of climate change.

Rising CO_2 levels in the atmosphere are a key factor in determining global warming. Climate change in recent years has exacerbated floods and droughts around the world. Crop failures are occurring with increasing frequency, due to lack of local rains, while diseases such as malaria are spreading into new areas. Ice and permafrost are melting at increasing rates, which may cause sea levels to rise. CO_2 has become more abundant in the atmosphere because fossil fuels have been burned at increasing rates during the twentieth century.

The Climate Change Convention of 1997 in Kyoto, Japan, established a system of 'carbon credits', in which the more trees a country plants in the ground, the more oil it could take out of it. To maintain their use of fossil fuels, therefore, industrial countries can seek to increase the planting of fast-growing trees within their own countries and elsewhere. In this way oil reserves can continue to be exploited with no change, without any 'carbon debts' being run up. Transgenic trees with enhanced growth rates may in the future increase the capacity of forests to act as carbon sinks. This is the reason why some key investors in transgenic trees have been automobile and oil corporations. The Japanese car manufacturer Toyota, for example, has conducted field trials with trees that have been genetically engineered for increased carbon sequestration and storage, although to date these trees have consumed unacceptably high quantities of water. British Petroleum (BP) and Shell are also funding work on transgenic trees in the hope of gaining 'carbon credits' (Owusu, 1999; Langelle, 2001).

World leaders at Kyoto pledged their countries to a 5 per cent cut in greenhouse gas emissions from 1990 levels by 2012. However, emissions have continued to rise by about 1.3 per cent annually. The USA has 5 per cent of the world's population, but accounts for more than 20 per cent of carbon emissions. The desire of the USA to exploit tree planting options, so as not to reduce its use of fossil fuels, led to the breakdown of the subsequent Climate Change Convention of 2000 in The Hague. European countries wanted to see more pro-active measures taken to reduce carbon emissions and were sceptical about the proposed tree-planting strategy (Warwick, 1999; Bunting, 2000). George W. Bush was quick to renege on even the limited US commitments made in Kyoto when he became president in November 2000.

The validity of the US strategy of planting forests rather than cutting carbon emissions has been called into question by scientific studies, such as those conducted in controlled-atmosphere environments in plots of loblolly pine (*Pinus taeda*) in a North Carolina forest. In these experiments, it was found that, although increased CO_2 produced an initial growth spurt, this rapidly slowed down because the trees soon depleted the nutrients in the soil. Further CO_2 could not be as effectively soaked up because of other limiting factors. Most forest soils have limited supplies of nutrients and other resources needed for growth. The amount of forest leaf litter also increased, and was broken down more quickly, in CO_2-enriched environments, releasing captured CO_2 back into the atmosphere. The overall effect was that the increased uptake of CO_2 by trees was short-lived. Therefore, there is probably only a limited potential for planting trees to act as carbon sinks for the rising CO_2 in the Earth's atmosphere (Schlesinger and Lichter, 2001; Oren *et al.*, 2001). The protection of existing forests and the expansion of the forested area are laudable goals, which will keep a significant amount of the planet's carbon stored as biomass. However, tree planting is no substitute for reducing carbon emissions when it comes to lowering the excessive CO_2 levels now present in the atmosphere.

In general, transgenic trees are likely to be economically beneficial to the large corporations harvesting timber and wood pulp from managed forests. Some environmental benefits may also occur. But it is unlikely that transgenic trees with faster growth rates will make either a measurable difference to the rate at which native forests are being lost or have beneficial effects in terms of climate change. Meanwhile, the longevity, size, and reproductive capacity of trees mean that there are a number of potential ecological risks. There is too little knowledge concerning these ecological risks to say that large-scale commercial plantings of transgenic trees will not have significant environmental impacts.

10 Fish

Aquaculture is the fastest-growing sector of the world food economy, with 31 million tonnes of fish being produced in 1998 compared to only 13 million tonnes in 1990. Production may double during the first decade of the twenty-first century. This is in contrast to fisheries production, which has remained static at around 92 million tonnes owing to the depletion of fish stocks by overfishing. Aquaculture may soon become the main means of fish protein production (FAO, 1997b). Fish farming is also catching up with cattle ranching, in terms of the number of people it feeds. Aquaculture needs relatively little land area, in contrast to beef production, is more productive in terms of weight gain per unit feed input, and even requires less water overall than needed to keep cattle alive (Brown, 2000a; Pimentel, 2001). Nevertheless, productivity is restricted by the prevalence of fish disease, while selective breeding for increased size and growth rates has been limited by existing gene pools. Genetic engineering could therefore play an important role in the genetic improvement of fish species in the future. This chapter describes how genetic engineering is being used to bring benefits to aquaculture, and the potential ecological risks that need to be addressed.

Modifications and benefits

Novel genes are introduced into fish eggs by microinjection, soon after they have been fertilized. As with other genetically modified organisms (GMOs), transgenes are integrated within gene constructs, which also contain promoters and marker genes (Iyengar et al., 1996). The marker genes most commonly used in transgenic fish have expressed the enzymes chloramphenicol acetyltransferase and ß-galactosidase, both of bacterial origin; firefly luciferase; and a green fluorescent protein from jellyfish. The promoters used in transgenic fish were originally of viral origin, but more recently tissue-specific promoters of fish origin have been used. There has been a move towards gene constructs containing fish genes whenever possible, preferably transgenes from the same or closely related species (autotransgenic fish). The number of gene constructs injected into fish eggs is usually large, and multiple copies can become integrated into fish genomes, at either one or several chromosomal locations. Transgene expression depends on positional effects (gene constructs integrate into fish genomes at random) and the number of copies expressed, but mainly on the effectiveness of the promoter. The techniques for integrating transgenes into fish are now well developed and described in numerous specialist books (e.g. Maclean and Rahman, 1994; Maclean, 1998).

A large number of genes have been integrated into fish genomes in order to change their physiology, behaviour, or tolerance of abiotic factors. The identification of genes conferring resistance to disease would be of major benefit to the aquaculture industry. It would also provide benefits to the environment, by cutting back on the large amounts of antibiotics and toxic chemicals administered to control fish lice and other diseases, and the number of diseased fish that escape to infect wild populations. However, disease-resistant transgenic fish are some way into the future. Meanwhile, a number of growth hormone genes have been isolated from fish and animals and expressed in transgenic fish.

The salmonids have been the subject of most research involving transgenic fish grown for food, in particular Atlantic salmon (*Salmo salar*), Pacific or coho salmon (*Oncorhynchus kisutch*), Chinook salmon (*Oncorhynchus tshawytcha*), and rainbow trout (*Oncorhynchus mykiss*) (Hew *et al.*, 1995). Most of these species are reared from eggs in onshore units and then transferred as young adults to large sea cages off the coasts of developed countries. Developing countries are also increasingly turning to aquaculture to provide a cheap source of protein. This aquaculture is more likely to be in freshwater pools, and may involve species of catfish, such as the channel catfish (*Ictalurus punctatus*) and the African catfish (*Clarius gariepinus*), or tilapia (*Oreochromis niloticus*). Herbivorous species, such as the common carp (*Cyprinus carpio*), can also be grown in rice paddies. Transgenic lines of all these freshwater fish have been produced (Dunham *et al.*, 1992; Horvath and Orban, 1995; Müller *et al.*, 1992; Rahman *et al.*, 1998).

Growth enhancement   Transgenic fish expressing growth hormone genes have increased growth rates and overall body size compared to unmodified fish. The final size of fish is less rigorously controlled by genetic factors than in mammals, and the potential for enhancement is considerable. A wide variation in growth enhancement has been reported, partly due to the number of different gene constructs used, with growth hormone genes and promoters from a variety of sources. Dramatic increases in growth rate and size have occurred. In experimental trials in the early 1990s, for example, Pacific salmon modified with a growth hormone gene had up to 40 times the circulating growth hormone levels of, and were on average 11 times as heavy as non-transgenic controls; some were up to 37 times as heavy (Devlin et al., 1994). Atlantic salmon modified with a Chinook salmon growth hormone gene also had significant growth enhancement, and reached sexual maturity in two years instead of the usual three (Du et al., 1992). In contrast, transgenic lines of other species, such as northern pike (Esox lucius) and gilthead bream (Sparus aurata), have shown little response to the introduction of growth hormone genes (Maclean, 1998).

Tilapia engineered with additional copies of their own growth hormone gene were almost twice as heavy as non-transgenics (Martinez et al., 1996). In another study, transgenic tilapia were three times as heavy, with ten to twelve times the levels of circulating growth hormone, as unmodified fish (Rahman et al., 1998). Transgenic carp, engineered with a growth hormone gene from rainbow trout, grew up to 40 times as fast as non-transgenics in one study, and had muscle with a 7.5 per cent higher protein content, a 13 per cent lower fat content, and a lower moisture content (Chatakondi et al., 1995). A 20 per cent increase in growth rate of transgenic carp, compared to non-transgenics, was observed in another study (Zhang et al., 1990). Shellfish have also been genetically engineered for enhanced growth under aquaculture systems. The growth rate of red abalone, for example, was increased by up to 40 per cent compared to wild-types (Gutrich and Whiteman, 1998).

Marketing Transgenic fish are already in the marketplace. In the USA, transgenic salmon that grow around ten times faster than normal fish have been sold in supermarkets since the late 1990s. These fish can reach around 3.7 metres (12 feet) in length and weigh 74.6 kilogrammes (200 lb). In the USA, there is no need to label these fish any differently from conventional fish, and so consumers have been unaware that they are consuming transgenic fish. Growth-enhanced transgenic fish may soon be on sale in Asia, where there is a very large potential market. Applications to market transgenic fish have been made worldwide. However, marketing in European nations and other countries where transgenic salmon would have to be labelled as a GM product could prove more problematic, because consumers have already rejected a range of other GM foods.

Freeze tolerance A major constraint in the sea cage farming of Atlantic salmon along the North Atlantic coast is that high mortality can occur when winter brings cold water and icebergs south. This species cannot tolerate ice-cold water. Arctic fish, on the other hand, are adapted to life in cold seawater. They synthesize anti-freeze proteins (AFP) in their livers, which bind to ice crystals as they begin to form, thereby stopping ice formation. An AFP from the winter flounder (*Pseudopleuronectes americanus*) has been engineered into Atlantic salmon, which normally lack AFPs. Incidentally, this same AFP gene has also been engineered into other GMOs, including experimental frost-tolerant tomatoes. The AFP gene is expressed in the salmon's liver and shows seasonal variation, with highest activity in the winter. However, only single copies of the transgene had been successfully integrated by the late 1990s, and this was insufficient to confer freezing tolerance in fish (Shears *et al.*, 1991; Hew *et al.*, 1999). If multiple copies of the AFP transgene can be integrated and expressed in Atlantic salmon, or a stronger promoter found, then they could be farmed in sites currently considered too cold for fish farming.

Physiological costs of growth enhancement

A balance must be struck between increased growth rate and the physiological costs of rapid growth. Unregulated over-production of growth hormone can be detrimental to health. Deformities, such as enlarged heads and a greater susceptibility to infection, for example, have been recorded in both transgenic Atlantic salmon and Pacific salmon modified for enhanced growth (Iyengar *et al.*, 1996). Deformities of this type have also been recorded in domesticated animals engineered with growth hormone genes, such as pigs (Pursel, 1997). It appears that a metabolic ceiling might be reached in all domesticated vertebrates, beyond which attempts to boost growth further by genetic engineering will result only in physical deformities.

In a study conducted by Robert Devlin and his colleagues in Canada, a slow-growing strain of wild rainbow trout could be engineered, with a salmon gene construct expressing growth hormone, to grow seventeen times as heavy and twice as long as its unmodified counterpart over a fourteen-month period. However, the same gene construct in a domesticated strain, which had already been selected for rapid growth by conventional breeding methods, had no effect on growth, but it did cause skull abnormalities and 100 per cent mortality before the fish reached maturity. In partly domesticated trout, growth hormone transgenes caused intermediate growth enhancement. The effect of introducing a growth hormone gene construct into fish appears to be dependent on the degree of earlier genetic improvement. Therefore, trying to enhance growth of domesticated fish and other animals might be harder to achieve than once thought, because artificial selection has already pushed metabolic systems towards their limit. In such cases, genetic engineering might provide limited benefits, while causing deformity and animal suffering (Ainsworth, 2001; Devlin *et al.*, 2001).

Growth enhancement might be highly beneficial, in terms of food production, for fish species having little history of domestication, particularly in the developing world. However, the costs of increased medical treatments, including antibiotics, and additional food requirements, need to be taken into account when assessing these benefits.

Risks to ecosystems

Environmental impacts of aquaculture   There are a number of well-documented environmental impacts associated with aquaculture, which may be exacerbated by its expansion, further intensification, and the introduction of transgenic fish. The main focus has been on farmed salmon and the primary concerns have arisen from the 'outputs', which include surplus food, fish waste products, fish lice and other parasites, medicinal products, and escaped fish capable of interbreeding with native populations. The nutritional content of water can be enhanced (eutrophication) by food pellets, which are administered automatically and fall through cages along with large quantities of ammonium-rich excreta. This can potentially change the ecology of an area, altering species abundance and diversity, via effects on energy balance and food chains. This may be particularly acute in sea lochs and other locations where tidal flushing is limited. A Canadian study has documented reductions in crustacea in lakes used for aquaculture. In contrast, there are a number of studies where effluent wastes from fish farms are diluted in marine environments and have minimal impact (Black, 1998). However, aquaculture is expanding and becoming more intensive, so its impacts are growing. The level of pollution from the fish farms on the west coast of Scotland has been compared to the sewage generated by a population of 9.4 million people (Millar, 2001). The eggs of fish lice released in sea cages can infest young migrating salmon (smolts) as they pass by on their way to the sea. The mortality this causes to wild salmon at sea may have contributed to a major decline in natural fish stocks in rivers in Scotland and Norway. The introduction of improved filter systems and 'no-loss' recirculation systems could help to reduce problems with waste products, but these are currently not used to any great extent.

Escape   Fish have proved difficult to contain in aquaculture and the number escaping is often surprisingly high. In Norway, for example, the

Directorate of Fisheries estimated that 500,000–700,000 of the country's farmed salmon escaped in 1994. In the USA, high levels of non-native escapee Atlantic salmon have been recorded from Pacific coast salmon farms. By the mid-1990s, around 35 per cent of salmon caught in coastal fisheries were escaped farmed stock (Charron, 1995). In some Norwegian rivers, escaped farm fish now account for more of the catch than wild fish. Fisherman can easily spot escaped fish because of their larger size, deformities (e.g. arched spines), diseases (e.g. cataracts), and their general poor quality compared to wild fish.

Escaped fish can have a considerable environmental impact. The accidental introduction of 21,000 juvenile pink salmon from a rearing unit into Lake Superior, for example, resulted in a population explosion two decades later that had widespread effects on the lake's ecosystem (Kwain and Lawrie, 1981). If transgenic fish with enhanced growth rates escaped containment they could consume more prey in rivers or exploit food resources not available to their non-modified counterparts, to the detriment of other species. The impact of transgenic fish might also cascade through food chains, increasing the probability of ecological disturbance (Gutrich and Whiteman, 1998).

The containment of genetically modified fish is therefore of great importance in reducing ecological risks (Devlin and Donaldson, 1992). Containment must be significantly improved if ecological impacts from transgenic fish are to be avoided. Even with better containment, however, it must be assumed that some transgenic fish will escape.

Persistence   In order for transgenic fish to have a major environmental impact they must not only escape containment, but also persist and establish viable populations. If transgenic fish become established, ecological risks could occur as a result of them posing a hazard in their own right or through interbreeding with the same or related species. Fish are not domesticated to the extent that livestock and crops are, and most species can readily survive in nature and interbreed with wild populations. Hybrids produced by the interbreeding of escaped farmed salmon and trout with wild species are common. Therefore, it is likely that transgenes will find themselves in wild populations soon after transgenic fish start escaping from sea cages.

The probability of transgenic fish, or hybrids containing their transgenes, persisting to establish populations depends on a number of factors, including what advantages a transgene confers, the number of fish that escape containment, the habitat into which they are introduced, the level of competition in this habitat, and the effectiveness of induced sterility or other methods of preventing escape and establishment.

Transgenic fish may be less fit than wild-types and therefore less likely to establish in the environment. Salmon modified with growth hormone genes, for example, were found to be lower in overall fitness and poorer swimmers than non-modified fish (Farrell *et al.*, 1996). Adverse physiological and behavioural changes have already been noted for fish with high growth rates. Some authors have argued that fitness reductions may be normal for transgenic fish (e.g. Maclean, 1998). However, transgenes for enhanced growth could in some cases lead to increased reproductive fitness. Juvenile mortality is usually high in fish, for instance, but enhanced growth rates could reduce mortality at this stage and enable more fish to reach reproductive age.

Concerns have been expressed for some time about the possible ecological impacts of growth-enhanced fish. An experiment with transgenic salmon on Loch Fyne in Scotland was abandoned in the early 1990s because of fears that fish engineered for large size might escape and interbreed with local fish populations (Brown, 2000b). Enhanced growth could make transgenic fish more competitive than unmodified wild-types, by enabling them to eat a larger range of prey items and thereby expanding their diet. This effectively changes the realized niche compared to the wild-type. Therefore, if transgenic fish with enhanced growth rates were to escape containment within their home range they would have a high probability of establishing in the face of competition from wild-types, with the result that wild populations could be reduced. However, outside of their home range and the conditions to which they are pre-adapted, they are much less likely to form viable populations.

An additional concern with the use of growth hormone transgenes is that they also facilitate adaptation to seawater in salmonids. This is not a problem in aquaculture, because farmed salmon are reared firstly in freshwater and then spend the rest of their lives in cages in seawater environments without returning to freshwater to breed. If growth hormone transgenes became established in wild populations, however, fish might become less well adapted to life in freshwater. In Atlantic salmon, for example, the ability to return to the river of their birth (using specific odour cues) might be lost. There has been an alarming decline in Atlantic salmon, due to river pollution and other factors. An additional factor could precipitate local extinctions.

Genetic improvements that enable organisms to overcome extreme ranges of abiotic factors, such as temperature, pH, or salinity, which restrict parental (non-modified) populations, enable modified organisms to exploit novel resources. GMOs modified for such changes therefore have modified fundamental niches. Although the AFP may not increase fitness within the parental range, it would enable a transgenic fish to push beyond the parental or native range. Salmon having AFP genes, for example, could exploit resources in coldwater areas unavailable to their parents. Any escape of freeze-tolerant transgenic fish could therefore present unique ecological risks (Gutrich and Whiteman, 1998). If the AFP transgene established in wild populations through interbreeding, for instance, the pattern of spawning in rivers could be dramatically altered, with implications for river fishing and river ecology.

The biggest ecological risks may come from transgenic fish that have both increased competitiveness in their native range, due to genes for enhanced growth, while also having a novel characteristic, such as freeze tolerance, that enables them to expand their range and exploit novel resources. Such fish would have expanded realized and fundamental niches. They would present a particularly high level of ecological risk because they could establish in their parental range to the detriment of wild populations, while having the ability to expand that range (Gutrich and Whiteman, 1998).

Sterilization Measures can be taken to reduce potential ecological impacts. Transgenic fish could be sterilized, for instance, either chemically or using genetic modification, before release into sea cages to prevent escapees from breeding with wild fish stocks. Transgenic sterile salmon are being developed in the laboratory with reversible sterility, so that they can breed when treated with a hormone they are engineered to lack (Charron, 1995). The primary developer of transgenic fish in the USA, A/F Protein of Massachusetts, claims the ecological risks are low because their fish are sterile. However, it will be impossible to achieve 100 per cent success with sterilization methods. Sterility in some cases might also adversely affect performance or behaviour of transgenic fish, leading to a drop in productivity. The release of sterilized fish would reduce but not completely eliminate long-term ecological risks (Kapuscinski and Hallerman, 1990; Knibb, 1997).

The need for transgenic fish

The substantial benefits that transgenic fish could provide means that the aquaculture industry is keen for research to move forward, particularly if it is aimed at producing large disease-free fish. However, salmonids are in danger of being overproduced, and there is resistance to GM food generally. Meanwhile, the intensification of aquaculture has thrown up food safety concerns. Farmed fish are increasingly being fed with industrial catches of small fish trawled from the seabed. These are concentrated into high-protein pellets. About a third of fish now caught commercially around the world go to feeding farmed fish or livestock. Oily fish of this type accumulate pollutants, which are passed on to farmed salmon. Salmon fed on pellet diets have been found to have high levels of PCBs (polychlorinated biphenyls), dioxins, organochlorines and other persistent and toxic man-made chemicals. Meanwhile, a mysterious brain disease was reported in farmed salmon in Scotland in the late 1990s. Given that recycled salmon are also used in fish pellets, the emergence of a salmon form of spongiform encephalopathy is a theoretical possibility.

Intensive fish-farming practices are due for an overhaul in the consumer's interest. What started as a cottage industry in the 1960s has become an intensive high-input means of food production, operated by multinational corporations such as Nutreco. Transgenic fish represent a further intensification of aquaculture, but is this what consumers really want?

Aquaculture could be made safer for the environment with better-sourced food, sea cages sited only in areas with adequate tidal flushing, new 'no-loss' systems to reduce outputs to the environment, many fewer escapees and the introduction of sterility measures. However, given the stress caused by high stocking densities, questions must be asked about the desirability of intensive production with transgenic fish. Indeed, given the known escape rates of farmed fish and the potential ecological risks, a moratorium should be put in place to prevent transgenic fish being released into sea cages.

The major benefits of transgenic fish may be derived in developing countries, where species such as tilapia and carp can provide additional protein that can answer real dietary needs. In landlocked and contained freshwater pool conditions, catfish and other species can be raised with far fewer ecological risks, because they are less likely to escape into natural habitats. A range of other fish with little history of domestication could also benefit from growth enhancement through genetic engineering. Meanwhile, tilapia are being genetically engineered to produce human insulin and other nutraceutical products, suggesting that transgenic fish could play an important role in aquaculture in the future. In general, however more attention needs to be paid to the potential ecological risks posed by transgenic fish, while comparisons with alternative approaches of increasing productivity should not be sidelined.

11 Assessing
ecological risk

A comprehensive evaluation of ecological risk should ideally be done before a genetically modified organism (GMO) is deliberately released into the environment. However, scientific knowledge is often lacking when it comes to the ecological risks associated with GMOs. In addition, a number of mechanisms at the molecular level complicate risk assessment. Field trials can provide data to help in assessing ecological risk, although they may create concerns in their own right. In this chapter, a number of factors that complicate risk assessment are described, the ecological risks associated with some additional types of GMO are considered, and the rationale behind the farm-scale evaluations in the UK is examined.

Risk assessment

Risk is the probability of an action having undesirable consequences. Risk assessment evaluates the probability of an action happening and its consequences (Wachbroit, 1991). In the case of releasing a GMO into the environment, risk is related to the probability of it escaping containment, the probability of it (or its transgenes) establishing outside of its release area, and the probability that it will have adverse ecological effects.

The release of a GMO into the environment can be considered as comprising a number of stages, to identify where risk lies. For example, the risk posed to honeybees by an insect-resistant transgenic crop can be considered in terms of pathway, receptor, effect, and hazard (DETR, 2000). The pathway is created when a bee visits a flower; reception is when a bee eats the pollen; the effect might be bee mortality, due to the hazard of the insecticidal toxin expressed in the pollen. If an element is missing from this sequence – for example, honeybees never visit those particular flowers or the toxin is not expressed in their pollen – then there is little or no overall risk. If pollen is also found on plants surrounding the transgenic crop in the field, the model can be adjusted accordingly. Therefore, even though mortality of honeybees, butterflies or other non-target species may be shown in laboratory studies, the full range of events in the field needs to be considered in risk assessment.

A framework for assessing the ecological risks of transgenic crops in the

USA was presented by Rissler and Mellon (1996). This was based on knowledge of the known weediness (invasiveness) of the parent (unmodified) crop, the presence of wild relatives, the frequency with which hybrids occur between wild relatives and the parent crop, and the likelihood of the genetic modification altering invasiveness or gene flow parameters. Transgenic crops could therefore be placed in different risk categories. However, if risks are unknown or unpredictable they cannot be anticipated by such a scheme. In addition, risk categories cannot be applied universally. An approval in the USA, for example, even one based on an assessment of ecological risk, does not ensure global safety.

The gene flow from a transgenic crop can be estimated in field experiments, in terms of potential gene flow (amounts of seed and pollen produced) or actual gene flow (counting marker genes transferred to receptor plants). Risk due to gene flow will decline with distance from a transgenic crop, past a point where it is considered negligible. If gene flow presents no mechanism for harm, then the logical response is to say that there is no overall risk. Therefore, by this reasoning, the spread of transgenes is of itself not enough reason to prevent the release of a GMO into the environment. Transgene flow must present a hazard before it can be said to have an environmental impact. Ideally, however, risk assessment should involve both biological and socio-economic criteria (Gliddon, 1999). In the case of transgenic oilseed rape, for example, it should take into account the ability of weeds and non-transgenic crops to acquire herbicide-resistant and other transgenes, but should also consider the economic impacts of genetic contamination on growers of organic or seed crops. The mere presence of nearby transgenic crops has adversely affected organic farmers in some European countries due to the adoption of a precautionary approach to those crops.

GMOs are regulated on the basis of their biological properties (phenotype) rather than their genetic make-up (genotype) or the genetic techniques used to produce them. This means that the same criteria can be used to evaluate transgenic and non-transgenic organisms, particularly those modified for the same characteristic. Risk assessment is important for all genetic improvement, whatever the means by which it is achieved. For example, risk assessment identified that a brown sorghum modified for bird-resistance through conventional breeding had high levels of tannin in its seeds, which also caused them to have anti-nutritional properties when eaten by humans. Herbicide-resistance has been obtained by both conventional breeding and genetic engineering. The environmental impact of herbicide-resistance may be due primarily to changes in crop management, rather than to any direct effects of the genetic modification process itself. Therefore, the characteristic of herbicide-resistance should be assessed similarly, irrespective of the means used to obtain it.

Nevertheless, the novel combinations of properties that GMOs possess suggest that they should be considered as distinct from organisms produced by conventional breeding techniques. Even those with the same characteristics, for example herbicide-resistant plants, differ in that the transgenic crops contain gene constructs with marker genes and promoters that are absent in conventional cultivars (Tiedje *et al.*, 1989). Ecological risk assessment should ideally be done on a case-by-case basis for GMOs being released into the environment for the first time, due to the numerous ecological, behavioural, physiological and molecular factors involved. Risks associated with marker genes and promoters have already been considered (Chapter 8). However, a number of other mechanisms at the molecular level also complicate the risk assessment of GMOs.

Transgenes within genomes

Gene silencing Gene constructs integrate themselves at random into genomes. The position in which they come to lie can be important in terms of transgene expression. Transgenes can be switched off at any time by the mechanism of gene silencing. All cells contain all an organism's genes, but not all genes are relevant to a particular cell. A gene that expresses a protein involved in plant root biochemistry, for example, will be silenced in the leaves by a process called methylation. Genes can be temporarily methylated within a cell. This allows them to be switched back on at a later stage of development. Transgenes can similarly be inactivated, although, due to positional effects, when and where this might occur is difficult to predict. The first case of transgene silencing was noticed in the late 1980s (Matzke *et al.*, 1989). Many instances have since been recorded, including loss of herbicide-resistance through transgene inactivation (Metz *et al.*, 1997).

Plant pathogens can initiate gene silencing. This reaction forms part of a plant's defence against pathogen attack, by restricting the development of infection. Transgenes can be inactivated as a side effect of this defence reaction. In transgenic oilseed rape, infecting viruses inactivated a transgene by disabling its promoter (CaMV 35S), which is of viral origin (Al-Kaff *et al.*, 1998; Senior and Dale, 1999). Transgenes that have been silenced, by whatever mechanism, could become reactivated at any time. Meanwhile, transgenes can become integrated within endogenous genes, owing to their random insertion in genomes. This can lead to the inactivation of other genes, causing unexpected effects.

Pleiotropy   Transgenes have been introduced in order to modify single characteristics, such as resistance to a particular insect pest. However, most genes can simultaneously affect more than one characteristic. These genes are said to be pleiotropic. The characteristics involved are often completely unrelated. In butterflies, for example, genes responsible for different variations of the colour brown also produce different resistances to temperature (Tudge, 1993). The mechanisms underlying such pleiotropic effects are poorly understood. Pleiotropic effects may arise from the insertion of transgenes. For example, a transgene might interact with an endogenous gene to produce an unforeseen effect involving an unrelated characteristic. Pleiotropic effects can also occur as a result of conventional plant breeding, but may be more unpredictable with novel gene constructs.

Gene stacking   The stacking of transgenes further complicates risk assessment. Commercially released transgenic crops increasingly have multiple transgenes. Stacked genes for herbicide-resistance and insect-resistance, for example, were integrated into 7 per cent of the transgenic cotton and maize (corn) grown worldwide in 2000, while crops with stacked transgenes are expected to gain an increasing share of the global market (James, 2000). The risks associated with each individual characteristic should be taken into account. In the future, additional characteristics, such as resistance to drought or saline conditions, are also likely to be stacked. This creates additional layers of complexity. Possible interactions between characteristics therefore have to be considered.

Transgenic crops of the 'first generation' have typically incorporated only one gene construct, but in the future many gene constructs may be added. Golden Rice (enhanced vitamin A) has three gene constructs, which carry genes for four enzymes plus the promoters and marker genes required to perform a multi-step biochemical transformation. As the number of transgenes and gene constructs increases in plants, the possibility of unpredictable effects occurring at the molecular level increases.

Transgenes can also become accidentally stacked in plants as a result of gene flow. The stacking of transgenes from independent sources can create particular risks. Weeds or crop plants that acquire herbicide-resistance transgenes from different crops, for example, can become resistant to a range of herbicides. This problem has already surfaced in canola in Canada (Chapter 6). Assessing the risks of gene flow to wild relatives becomes more complicated when transgenes for different characteristics are involved. The consequences of gene stacking should therefore be carefully considered in risk assessment.

There has been a tendency to view genetic modification in an overly reductionist manner. Genes can no longer be viewed merely as discrete, independent units arranged linearly on chromosomes. Genes interact in complex networks. Holistic models of gene function are emerging that offer better explanations of gene function. These account for the robust nature of genetic architecture, but also the often unpredictable nature of genetic modification. For example, if genetic engineers 'knock out' single genes, the effect will always be predictable using a reductionist model, but in reality knocking out single genes often fails to have any noticeable effect. Single gene changes can shift a plant's biochemistry in unexpected directions because of compensatory effects acting throughout the network of interacting genes. The holistic approach, which is more useful in helping to understand the unpredictable nature of genetic change, can be seen to be at odds with both the practice of genetic engineering, in which the effects of only one or two genes are considered, and its popular image, as advanced by the biotechnology industry (Ho, 1998).

Risk assessment will vary with the type of GMO being released to the environment. The risks associated with a range of transgenic organisms not featured elsewhere in this book will be briefly considered here.

Arthropods, birds and mammals

Transgenic arthropods   The first field release of a transgenic arthropod occurred in Florida in 1996. The beast involved was a predatory mite (*Metaseiulus occidentalis*) with a marker transgene, which was tracked within a fenced area. The aim was to study the persistence of the transgene in a field population for risk assessment purposes. The marker gene was rapidly lost when transgenic individuals interbred with a wild colony (Hoy, 2000).

Insects have been genetically engineered primarily within laboratory-based research programmes, but some are being developed for future deployment in the field. They may be released as part of pest management strategies, such as the sterile insect technique (SIT). The mass release of sterile insects was first successfully used to control screwworm (*Cochliomya hominivorax*) in the USA about 50 years ago, and has now been applied against about 30 insect species. The method is effective because the large number of sterile males released mate with wild females, in competition with wild (fertile) males, reducing the number of fertilizations and causing populations to crash. Males are usually made sterile in the laboratory with radiation or chemicals that induce mutations in the sperm. The mass release of transgenic insects extends the possibilities of the SIT.

Sterile pink bollworms   Transgenic pink bollworm (*Pectinophora gossypiella*) moths, whose caterpillars are a major pest of cotton in the southern USA, were first released in field cages in Arizona in summer 2001. These insects contained a marker gene, expressing a green fluorescence protein from a jellyfish (*Aequoria victoria*), and a promoter from a fruit fly (*Drosophila melanogaster*). The marker gene enables quick and easy screening of mass-released moths in the field. In releases from 2002 onwards, the moths will also be engineered with a transgene that causes sterility. This will be cheaper than inducing sterility by methods previously used in the SIT, while transgenic insects are expected to be less physiologically damaged than irradiated ones (Brasher, 2001).

One approach being developed in the laboratory with pink bollworm is to engineer them with a fruit fly gene that expresses a protein damaging to insect metabolism. The laboratory-reared moths, however, will be treated with a chemical that acts as an antidote. In the field, male moths compete with wild moths as in the classic SIT, but, instead of being sterile, pass on the toxin gene. The offspring of females have no access to the chemical antidote and will die. The large number of transgenic insects released, the possibility that the transgene will spread to other species, and possible indirect effects on non-target species will have to be considered in risk assessments of such releases (Meek, 2001).

There has been a range of proposed applications involving transgenic insects. For example, transgenic mosquitoes are under development that could deliver vaccines or carry their own antimalaria genes. However, the ecological risks associated with the release of transgenic arthropods could be relatively high, due to their high mobility, short generation times, and high reproductive rates (Hoy, 2000).

Birds   As with other GMOs, the ecological risks associated with escaped transgenic birds and animals will partly depend on their ability to establish in the environment, and on how much their modification affects their natural ecological role. Relatively few genes have been isolated from birds, which has limited the options for genetic engineering. Most agricultural research on transgenic birds has been with chickens, with the aim of raising productivity in battery units. Transgenic chickens are being designed to have either increased growth rates or resistance to bacterial infections prevalent in cramped rearing conditions, such as *Salmonella*. Escaped birds of this kind have lost most of their wild-type characteristics and are not adapted to an undomesticated life. They would therefore have a low probability of establishing populations outside rearing units (Simkiss,

1994). The main concerns with transgenic chickens relate to welfare and the further intensification of production.

Mammals Mice account for around 98 per cent of all transgenic mammals. They are mainly used in experiments on human disease, are held in secure laboratories, and are poorly adapted to life in the wild. Other laboratory mammals also have a low probability of escape under normal circumstances. However, transgenic mammals are being developed for life on the farm. Pigs and cows have been modified in experiments for increased size, using transgenes expressing growth hormones. These animals have tended to be more prone to illness than their unmodified siblings, which raises significant welfare issues. They would probably not survive long if they escaped from farms. Some domesticated animals, such as rabbits and goats, do have a history of establishing populations in the wild, however, and genetic exchange could occur between domesticated animals that escape and wild populations. Wild boars have been known to break into pig rearing units and mate with domestic sows, producing hybrid piglets. These animals present a higher level of ecological risk, dependent on the transgenes involved.

Transgenic clones of sheep, cows and goats, modified to produce human therapeutic proteins in their milk, may soon be released into fields. These animals could shortly be providing cost-effective treatments for emphysema, cystic fibrosis, and a form of haemophilia. A range of other human proteins could also be produced in this manner. However, individual 'pharmed' animals will be worth a great deal of money and are likely to be closely monitored (Brem and Müller, 1994). Further in the future, transgenic animals could be made resistant to a range of diseases, for example scrapie in sheep, or foot and mouth in cattle. Such animals could provide considerable benefits, although transgenic solutions do nothing to cure the root causes of animal disease. The main concerns with transgenic animals are ultimately nothing to do with them escaping into the wild, but with a range of social and economic issues. These include animal welfare, patenting and the control of agricultural production, and the loss of genetic diversity.

Farm-scale evaluations

Experimental trials of transgenic crops have been numerous. However, in the majority of cases plot size has been small and ecological questions have not been asked. Industry trials are typically done on an area the size of a tennis court, in order to obtain data on yields and crop performance. In

Britain, plots of 40 m² have been used to obtain data for National Seed List Registration. From the information obtained from small-scale trials, transgenic crop seed can be approved for marketing. In the USA, companies are under no obligation to collect ecological data from field trials. Small-plot trials have been followed by cultivation in fields that extend as far as the eye can see in mid-western states.

The development of transgenic crops has been market-led. There has been a lack of cooperation between industry and independent ecologists, while few combined agronomic and ecological field studies have been conducted. Therefore, little ecological data is available for assessing the risks posed by transgenic crops in the field. Vague statements about negative effects, such as 'no effects on non-target organisms' or 'no increased weediness', that have arisen from many trials have insufficient ecological data to back them up. Ecological studies that use non-transgenic crops to predict the population dynamics of transgenic plants, meanwhile, lack the necessary relevance or predictive power (Snow and Morán Palma, 1997; Stewart and Wheaton, 2001).

Ecological studies with transgenic crops, conducted on a commercial scale and combined with agronomic evaluation, are therefore desirable. These could identify ecological risks, which may lead to the safer long-term cultivation of transgenic crops, but such experiments also have their own environmental impacts. This is the dilemma facing countries that wish to proceed with transgenic crop technology, but take a precautionary approach.

In an extreme case, an experimental trial of transgenic maize was completely swept away by flooding in Iowa in 1993. A moat, a fence, and a vegetation-free area surrounded the riverside trial, but nothing could prevent it being uprooted and carried downstream in a torrent. The plants were never recovered (Anderson, 1993). In less extreme cases, some level of gene flow is likely to occur from an experimental trial to wild relatives or non-transgenic crop cultivars. Nevertheless, measures can be taken to reduce the risks associated with experimental trials. If transgenic crops are to be cultivated over a large area, then having data from ecological studies is better than having no data.

Farm-scale evaluations are not a regulatory requirement before commercial release of transgenic crops in industrialized countries or in the developing world. Few agronomic or ecological studies have been conducted in developing countries, particularly those with wet or arid tropical climates, although areas within these countries have been proposed as sites for transgenic crop cultivation. The results from experiments in temperate climates cannot just be extrapolated and considered relevant to these areas. Agronomic performance might be very different, while levels of invasiveness or transgene flow might be higher in the tropics, particularly in regions of high biodiversity.

The number of ecological studies conducted also varies for developed countries. In the UK, for example, there has been ten times more funding for scaled-up risk assessment and agronomic experiments than in the USA, even though the crops in question were developed in the USA (Stewart and Wheaton, 2001). Other European countries have also conducted a range of ecological studies with GMOs. (Many of these studies have been referred to previously, e.g. Chapter 6). The UK farm-scale evaluations will be examined here to illustrate the work involved, the problems encountered, and the limitations inherent in an ecological study of this type.

UK trials   The farm-scale evaluations in the UK, announced in October 1998, were set up partly in response to public concerns about the environmental impacts of transgenic crops. The government-funded trials were accompanied by a voluntary agreement with industry, represented by the umbrella organization SCIMAC (Supply Chain Initiative on Modified Agricultural Crops), that no commercial planting of transgenic crops would occur until the three-year evaluation programme had been completed (DETR, 1999a). A Consortium was commissioned to conduct the evaluations, consisting mainly of scientists from government agricultural research institutes, while an independent Scientific Steering Committee has overseen them, advising on experimental protocol and publication. Preliminary trials were conducted in 1999, to fine-tune the methodology, followed by three years of full-scale evaluations, which started in 2000. The results are likely to be made public in 2003 (DETR, 1999b).

In Britain, most proposed transgenic crop releases have been for herbicide-resistant lines. This modification was chosen therefore for the four trial crops: spring and winter oilseed rape, fodder maize, and sugar beet. The evaluations were designed to look for differences in farmland biodiversity between genetically engineered herbicide-resistant crops and comparable conventional crops. Biodiversity was defined as the range of variation in living creatures, at genetic, population, and community levels. This was quantified using data on the relative abundance of selected animal and plants. The main question asked in the study was whether there are any differences in biodiversity between herbicide-resistant and conventional crops. The design tested for both positive and negative effects. It was understood that any effects were likely to be indirect, resulting from changed crop management. The trials therefore were not about the biosafety of transgenic technology *per se*. Herbicide-resistance is a characteristic that has also been introduced into crops by conventional breeding, and the experimental design would have been the same if these cultivars had been used (Firbank *et al.*, 1999).

A split-field experimental design was used in the farm-scale evaluations, with 10 hectare (25 acres) planted to the herbicide-resistant treatment and the rest planted with an equivalent conventional crop (Firbank, 1999). No baseline data was collected on the farm sites, which critics saw as a weakness in the experimental design. In split-field comparisons baseline data is not strictly necessary, but a survey of wildlife beforehand would have given a fuller picture of the effects of transgenic crops on the environment and biodiversity. For statistical purposes, a minimum of 12 to 15 farms was required per crop. The aim was for 25 field sites of fodder maize and oilseed rape, and 30 fields of sugar beet, to be planted in each of the three years (Firbank, 2000).

Biodiversity is influenced by numerous factors, and therefore sites were chosen to represent a range of geographic variations in soil, climate, species occurrence, and farming intensities that would be encountered if transgenic crops were grown commercially. Instead of doing whole population assessments, relative counts of sensitive indicator species were used to assess biodiversity. The indicator species chosen were: slugs and snails; springtails (Collembola); ground beetles and other ground-dwelling arthropods; bugs (Heteroptera) and butterfly, moth and sawfly larvae on foliage; and adult bees and butterflies. Earthworm sampling was initially included in the protocol, but it was removed from the experimental design after the preliminary trials. Ground-dwelling invertebrates were collected with the aid of pitfall traps, which are open canisters sunk into the ground. Vortis suction samplers, which are basically vacuum cleaners consisting of a backpack and a wide nozzle, were operated by workers moving through a field to collect arthropods on vegetation. Total invertebrate biomass was estimated by weighing sub-samples and scaling up. Vegetation in and around fields, for example weed seedlings, was monitored using standard transect (line) and quadrat (area) techniques. Soil samples were analysed to obtain seed bank counts. The full sampling programme was performed three or four times annually at each site. Birds and small animals were not recorded directly because they range too widely for single field studies, although they were looked at in a separate study in 2000. However, the food resources of birds were measured, enabling the effects on birds to be indirectly assessed via the food chain (Firbank, 2000).

The Consortium had responsibility for the scientific conduct of the crop evaluations, but the responsibility for the supply of seed, the initial choice of farm sites, and recommendations concerning crop husbandry rested with SCIMAC. The Consortium made the final selection of farm sites, using scientific criteria, from among those chosen by SCIMAC. However, with the difficulty of finding farmers willing to volunteer for the trials, less choice was available than initially envisaged. Aventis CropScience UK produced most of the transgenic seed used in the trials. The oilseed rape and fodder maize were hybrid lines developed by Aventis, genetically modified to be resistant to Liberty®, a glufosinate herbicide. The sugar beet was tolerant of Monsanto's glyphosate herbicide Roundup Biactive®.

The job of conducting the farm-scale evaluations has been made difficult by pressure put on farmers to withdraw and vandalism by protestors. Of the seven sites in the preliminary trials of 1999 one oilseed rape site was taken out of the experiment by a farmer (Hannington, Wiltshire), while extensive damage was done to oilseed rape (Watlington, Oxfordshire) and maize (Lyng, Norfolk) sites. In the case of the Hannington trial, on Lushill Farm near Swindon, Captain Fred Barker ploughed in the transgenic oilseed rape nine weeks after it was planted, because the Soil Association threatened to withdraw organic certification for the 100 hectares (250 acres) of organic crops that were also grown on his farm. The Soil Association had initially advised the farmer that his organic status would not be threatened, but later toughened its stance when its director Patrick Holden announced that organic farmers would no longer be certified if they grew organic produce within six miles (ten kilometres) of transgenic crops. The Soil Association said that this was to avoid 'genetic pollution', but Aventis saw it as a method of persuading farmers not to participate in the trials (Arthur, 1999). Farmer withdrawals have continued during the farm-scale evaluations (e.g. Meikle, 2000d).

In the incident at Lyng (Walnut Tree Farm), Norfolk in June 1999, a dawn attack on a trial of transgenic maize resulted in Greenpeace's executive director (now retired) Lord Melchett and 27 other activists being arrested. The attack was organized after Greenpeace learned, from an article written in *Farmer's Weekly* by one of the farmers, that the maize would be flowering and releasing pollen by the end of June (Vidal, 1999). In Watlington (Model Farm), Oxfordshire in July 1999, around 100 people trampled a preliminary trial of transgenic oilseed rape, with six arrests being made (Quinn, 1999). The popular BBC radio soap 'The Archers', about a fictional rural community, typically reflected the *zeitgeist* when young Tommy Archer appeared in court for trashing a trial of transgenic oilseed rape. He was acquitted by a jury won over by one of its members arguing the case against transgenic crops (Leapman, 1999). This reflected the real-life situation in which sympathetic juries have acquitted many protestors of criminal damage to crops. 'The Archers' engaged a

number of advisors, including Monsanto, who provided a case for transgenic crops, and Mike Schwarz, a solicitor who has acted for environmental
campaigners. However, Schwarz had to withdraw from the advisory team
when Lord Melchett hired his services in real life. Melchett and his co-
defendants were acquitted of theft in a first trial in April 2000, and then
acquitted of criminal damage in a retrial in September. Greenpeace
successfully used a public interest defence, endorsed by juries who did not
regard direct action against transgenic crop trials a criminal offence.
Needless to say, industry groups, the National Farmers' Union (NFU) and
the government feared that such verdicts would give a green light to
vandalism and trespass (Kelso, 2000). In 2001, the level of damage to evaluation crops threatened the validity of the data being collected.

Those resisting transgenic crops in Britain see themselves as part of a
long tradition of rural dissent. Activists who destroyed a maize trial in
Devon made this explicit by comparing themselves to the Tolpuddle
Martyrs – union organizers who were arrested in Devon in 1834 and
deported to Australia, only to be pardoned in 1836 after massive public
support for their cause (Gibbs, 1998). Democracy has been bypassed in
the case of unpopular transgenic crops and GM food – or at least that is
a widespread perception – so protestors consider illegal acts to be justified,
or even a civic duty (Monbiot, 1999). However, when demonstrators
destroy studies that could contribute to an understanding of the environmental impacts of transgenic crops, they risk undermining claims that they
are seeking a rational debate on the issues.

Environmentalists have raised a number of objections to the farm-scale
evaluations. These were collated by Friends of the Earth, under the sub-
headings 'creeping commercialization, environmental hazard, policy
chaos, and scientific farce' (FOE, 1999b). Under the charge of creeping
commercialization, it was pointed out that industry has pushed ahead with
additional experimental trials in order to register transgenic lines on the
National Seed List in preparation for commercial release. Therefore,
although no transgenic crops would go into the food chain during the trials
through voluntary agreement, things were being put in place for an
immediate large-scale release of transgenic crops, should the government
give the green light in 2003. The trials were therefore seen as commercially driven plantings with environmental studies tagged on. However, the
fact that the trials were to be done in fields as if they were commercial
plantings was always going to constrain the type of research that could be
done. The 'managed development' of transgenic crops was implicit in the
agreement made between government and industry, and it is understand-
able that industry be prepared to move ahead quickly if given the all-clear.

The charge that the farm-scale trials are an environmental hazard in themselves has been the most difficult objection to refute. The trials were not designed to prevent gene flow and therefore 'genetic pollution' was inevitable. By early 2000, the government accepted that the trials would cause a limited amount of contamination in conventional crops growing nearby. Organic growers and beekeepers stood to be most affected (Chapter 6). Cross-pollination could also occur with wild relatives, in the case of oilseed rape and sugar beet. To try and allay public fears, the separation distances between transgenic and conventional crops were increased in August 2000, from 50 metres to 100 metres (80 metres for maize), or 200 metres in the case of adjacent organic and seed crops. Government and industry calculated that these separation distances would produce genetic contamination levels of 1 per cent or less, and thus be within limits set by EU law. The separation distances, however, are clearly inadequate as a measure for preventing gene flow and some researchers have suggested that contamination could easily reach 5 per cent or more in nearby crops. Organic farmers have asked for separation distances of at least three kilometres. Meanwhile, beekeepers were advised to move their hives to around ten kilometres (six miles) from the nearest trial, which may also affect those farmers who rely on hive bees for pollination (Chapter 7). A leaked memo from the Cabinet Office in 1999 revealed a lack of clear policy on this issue with unanswered questions about how much environmental damage could be tolerated before a halt was called to the trials, and who would be liable for any damage caused. FOE argued that legal responsibilities should have been worked out prior to the trials (FOE, 1999b).

FOE's accusation that the trials are a scientific farce has unnecessarily annoyed scientists involved with the evaluations. In scientific terms, the trials are competent in design, and are conducted by professional scientists who, by and large, pride themselves on their independence. However, it is important that the limitations of such studies are understood. They are asking a very specific question, and cannot be expected to provide complete answers to questions regarding the biosafety of transgenic crop cultivation in Britain. In this respect, the British government has put a disproportionate emphasis on the trials within the decision-making process regarding transgenic crops. The debate needs to be cast in much broader terms. Ultimately, the decision will be a political rather than scientific one. Evidence of adverse effects on biodiversity would give the government the legal means to ban transgenic crops using the precautionary principle if they chose to do so (see Chapter 13). However, if the government decides to go ahead with transgenic crops and there is evidence of adverse ecological effects, measures to preserve biodiversity, such as leaving head-lands and crop margins unsprayed, could be implemented. In September 2001, the government's recently established Agriculture and Environment Biotechnology Committee concluded that the farm-scale evaluations by themselves would not provide enough evidence on which to base a decision about the commercialization of transgenic crops in Britain.

A key factor in the farm-scale evaluation design is how often herbicide is sprayed. High levels of spraying, which remove all weeds from herbicide-resistant crops with the aim of maximizing yields, would be most detrimental to wildlife. A regime of well-timed sprays, however, might be beneficial to wildlife compared to some conventional spray regimes. The written advice of the industry body SCIMAC was aimed at ensuring 'cost effective weed management', with herbicides being used as they would in commercial situations. Negative effects on biodiversity due to overspraying were unlikely in the farm-scale trials, which compared herbicide-resistant crops to regular high-input commercial crops, of the type already impli-cated in declining farmland wildlife. An average of five sprays are applied to sugar beet in conventional systems, for example, while a maximum of three herbicide sprays are recommended on transgenic sugar beet. Given the criteria used, industry could be fairly confident that differences between the treatments were never going to be great. Things might have been different if organic or low-input agricultural systems were included in the experimental design. Oilseed rape and sugar beet are not grown organically in Britain, but a small amount of maize is. It was explained by the project designers that organic maize was not considered as a treatment owing to the separation distance required between transgenic and organic crops, which would have prohibited a paired comparison (DETR, 1999c). The Scientific Steering Committee did emphasize the need to represent less intensive production systems beyond the first year of the study, but this stopped short of organic production. There remains a need for a real com-parison between transgenic and low-input production methods.

European trials   Farm-scale evaluations have been conducted elsewhere in Europe. In France, for example, a more modest multi-year and multi-crop experiment has been in progress since 1995. In three different regions of France, six-hectare fields of transgenic herbicide-resistant maize, oilseed rape, and sugar beet have been compared with conventional cropping systems. A report of the preliminary results concluded that the develop-ment of multiple resistance to herbicides, via gene flow between crop cultivars resistant to different herbicide groups, was the main concern for farmers, and of more significance than gene flow to weeds. The report suggested that agronomic practice be adjusted, so as not to jeopardise the sustainable use of transgenic herbicide-resistant cultivars, and that a more integrated management approach be implemented (Champolivier *et al.*, 1999). Farm-scale evaluations of transgenic maize in Germany, meanwhile, were abandoned in February 2001 due to growing public concern about food safety.

In Japan, where there has also been a low level of public confidence in GM food, a government-funded project was initiated in 1999 to look at the long-term environmental impacts of herbicide-resistant and insect-resistant transgenic crops. Every transgenic crop in Japan now has to be cultivated for at least one generation in a farm-scale study before it can be commercially released.

There are an increasing number of farm-scale studies in progress to aid in the risk assessment of transgenic crops. However, they are coming at a late stage in the development cycle of these crops, which are already cultivated over large areas around the world. Problems, particularly relating to 'genetic pollution', have therefore only become apparent in commercial situations. If, say, the extent of gene flow in canola had been realized earlier, would it have affected the rate of adoption of transgenic crops?

12 Diversity

Genetic engineering has been accompanied by declining diversity in a number of areas. A limited number of transgenic crops have been marketed, having a low genetic variability, for cultivation in monocultures around the world. The high-input requirements of transgenic crops, particularly herbicide-resistant ones, are likely to exacerbate the loss of biodiversity in and around farmland. In addition, a series of corporate mergers has reduced diversity in the commercial world, limiting choice to growers and consumers. There is a danger that powerful commercial interests may promote transgenic crops at the expense of a range of more appropriate and sustainable alternatives, particularly in the developing world. This chapter examines the various ways in which agricultural applications of genetic engineering threaten to reduce diversity.

Crop diversity

Although genetic engineering could be used to create a greater diversity of cultivated plants, the initial trend has been towards a narrow range of releases for the international market. Transgenic crops released to date have been designed for large-scale systems of production. They take after the Green Revolution varieties before them, in typically requiring high inputs of water, fertilizer and pesticides. Their cultivation may lead to the continuous cropping of monocultures, and a decreased use of rotation and polycultures. This simplification of cropping systems reduces crop diversity.

Monoculture vulnerability Modern high-yielding varieties have been bred from a narrow genetic base. The wheat hybrids released worldwide from the International Centre for the Improvement of Maize and Wheat (CIMMYT) in Mexico during the Green Revolution, for example, were derived from just three crosses of Japanese dwarf wheat and American wheat. These hybrids had shorter and firmer stems, which could support bigger heads of wheat. In the 1980s in the USA, around 70 per cent of maize (corn) was planted with just six cultivars. A single potato variety still covers around 80 per cent of the potato-growing area in The Netherlands.

The pattern is repeated throughout the industrialized world. Transgenic crops may be even less diverse than the high-yielding varieties they are replacing (Fowler and Mooney, 1990; Shiva, 1993).

Uniformity has been the aim in transgenic crop development. However, if a few transgenic crops replace a multiplicity of locally adapted crops, then food security could be compromised. Adequate food security is the access by all people at all times to enough food for an active and healthy life (Conway, 1997). Crop failures can result in inadequate food being available. Monocultures of single varieties are particularly vulnerable to pests and diseases. This was the case, for example, with potato blight in Ireland in 1845, 'red rust' in wheat in the USA in 1954, and corn leaf blight in maize in the USA in 1970.

Mosaics of plants having different genetics tend not to be devastated to such an extent by pests and diseases. A study of traditional 'sticky' rice varieties in China and the Philippines showed that when a number of them were planted side by side in the same field, a fungal disease (rice blast) was 94 per cent less severe and yields 89 per cent higher than in a uniform planting of one rice variety. Sticky rices are susceptible to rice blast, the major disease of rice, but they benefited from being planted amongst more resistant varieties. The use of intraspecific crop diversification in this case removed the need for fungicide sprays (Zhu *et al.*, 2000).

Reliance on a few crops   The number of widely cultivated plant species has diminished in recent years. Around 50 per cent of the world's calories come from just four crops (rice, maize, wheat and potato), while over 95 per cent of calories come from around 30 crops. During the Green Revolution this process was intensified, as high-yielding varieties of major crops replaced a range of locally adapted plants. Transgenic crops are part of a continuation of this trend. However, there are thousands of promising food plants worldwide. For example, India has around 1,000 edible plants, Ghana has around 2,500, and the arid Sahel region of Africa around 800. Many more can be found in South America.

A diversity of crops ensures that pests and diseases do not wipe out a large proportion of the food supply. A range of minor crops guarantees better food security than monocultures that are often grown for quantity rather than quality. Minor crops have suffered due to the process of globalization, which has seen them replaced by high-yielding, but not necessarily more nutritious crop species. Yet minor crops (which it will never be cost-effective to genetically engineer) are usually adapted to local conditions, especially niches in marginal areas, and may play an important role in food security during lean years (Pearce and Padulosi, 2000).

Crops and weeds   Cultivated plants and weeds are closer genetically and functionally than is often realized. Weed races have evolved from almost all of today's economically important cultivated plants, while many cultivated crops have directly evolved from weeds. In both cases, weeds can cross-breed with related cultivars. Gene flow is commonplace between crops and weeds, as previously noted. The similarity of crops and weeds can mean that the distinction between them is blurred. Species used for agricultural purposes in one part of the world may be considered weeds in another. In a listing of the 18 'world's worst weeds', 17 are also cultivated somewhere in the world (e.g. *Amaranthus hybridus*, *Avena fatua*, *Chenopodium album*, *Cynodon dactylon*, *Sorghum halepense*) (Bartsch *et al.*, 1993; Holm *et al.*, 1977). In traditional agricultural systems that embrace diversity, 'marginal crops' growing within or around fields of other crops can provide useful sources of fodder or vitamin-rich leaf vegetables, natural bioinsecticides, or medicines. These plants are considered weeds in monocultures of high-yielding varieties and are killed with herbicides. But marginal crops can provide all-important 'famine foods'. Monocultures therefore reduce local biodiversity, with an accompanying erosion of local knowledge systems, and may undermine food security (Shiva, 1993).

Declining crop diversity   The major centres of crop diversity are in Latin America, the Sino-Japanese region, Central Asia, and North-East Africa. These regions have the greatest genetic resources for major crops. The world's poorest nations therefore account for some 95.7 per cent of the genetic resources, while regions such as North America are totally dependent on external sources of genetic resources for their major agricultural crops (Juma, 1989).

During the twentieth century, as much as 75 per cent of the genetic diversity of agricultural crops has disappeared. As early as the 1920s, it was noted that traditional or indigenous crop varieties, known as landraces, were being lost from fields and smallholdings worldwide. Landraces are useful to plant breeders because they have considerable genetic variability, and are adapted to stressed environments and low-input agricultural systems. In the developed world many fruit and vegetable varieties are being lost; a situation that has been exacerbated by the closure and take-over of small independent seed companies. In the USA, for example, 80–90 per cent of fruit and vegetable varieties were lost during the twentieth century, including 85 per cent of apple varieties (6,121 in number), 88 per cent of pear varieties (2,354), and 95 per cent of garden pea varieties (546). The situation has been just as pronounced in the developing world since the Green Revolution. In India, over 30,000 different rice landraces have traditionally been grown, but a rapid decline has set in with at most ten varieties now cultivated on three-quarters of the rice-growing area. A similar trend has also occurred in livestock worldwide. In Europe, for example, over 750 breeds of domestic animal became extinct during the twentieth century (Fowler and Mooney, 1990).

On-farm conservation   The focus for conserving genetic resources from the 1930s to the 1980s was gene banks, which are centres where seeds are preserved in cold storage. However, since the 1990s more attention has been paid to the large resources of uncollected but conserved diversity on small farms. The importance of *in situ* conservation, particularly within the centres of crop diversity in the developing world, was acknowledged by the Convention on Biological Diversity in 1992. *In situ* conservation maintains genetic diversity within living viable ecosystems. It is a dynamic conservation process, which generates new genetic resources, for example due to gene flow between populations. Farmers are encouraged to select and manage local crops, and local knowledge is preserved as well as genetic resources. *In situ* conservation therefore provides a backup to collection, but captures key elements of crop diversity that cannot be stored in a gene bank. The maintenance of traditional farming systems is the best way of promoting on-farm conservation (Brush, 2000).

The deployment of sterile transgenic crop seed prevents farmers from saving, sharing and replanting seed. The selection of seed from the best plants in traditional plant breeding leads to the selection of crops that are adapted to local conditions. Farms practising seed saving in this manner act to conserve important crop diversity on-farm. If Terminator technology becomes ubiquitous in seed supplied from a centralized source, then local crop diversity cannot evolve. Furthermore, gene banks and *in situ* genetic resources are now at risk of being contaminated by transgenes from transgenic crops, especially in their centres of origin. Wild maize varieties have already been contaminated in Mexico.

Corporate diversity and the patenting of life

Corporate mergers   Genetic engineering has been accompanied by a decreased diversity in the commercial world. Mergers occurred at an unparalleled rate in the last years of the twentieth century, with a sudden enormous concentration of corporate power. The formation of Syngenta in 2000, when Novartis merged with AstraZeneca, was just one in a series of mergers that has seen the agribusiness multinationals getting bigger and fewer. Previously, a merger between Ciba-Geigy and Sandoz had formed Novartis, while AstraZeneca was formed from a merger between Zeneca and Astra. Syngenta became the world's largest agribusiness as a result of this merger, taking the accolade from Aventis (formerly AgrEvo), created the previous year through a merger between Hoechst and Rhône–Poulenc. Du Pont bought out the world's largest seed company, Pioneer Hi-Bred, in 1999. Monsanto kept up by merging with Pharmacia and Upjohn (Pharmacia) in 2000. The boundaries between biotechnology, agriculture, agrochemicals and pharmaceuticals have become increasingly blurred with the creation of these giant 'life science' corporations.

In the early 1980s, 65 companies competed in the global pesticide marketplace; by the late 1990s, less than ten companies accounted for over 90 per cent of total pesticide sales. The top four life science corporations (Syngenta, Aventis, Monsanto and Du Pont) now account for around 60 per cent of global pesticide sales, almost 23 per cent of commercial seed sales, and virtually 100 per cent of transgenic seed sales. Monsanto's transgenic seed accounted for around 88 per cent of total transgenic seed sales in 1998. Genetic engineering is very much a corporate science. The ongoing process of consolidation has reduced competition and choice in the marketplace. A small number of transgenic crop lines are being aggressively marketed, while little interest has been shown in increasing the genetic diversity of crops. Just two genetic traits (herbicide-resistance and *Bt* insect resistance) accounted for the vast majority of transgenic crops grown commercially in 2000 (RAFI, 1999; Shand, 2001).

Patent protection A transgenic world will be a world regulated by patents. Previously, Plant Breeders' Rights (PBRs) safeguarded seed company interests by giving them a monopoly on selling and marketing, but without giving them rights of ownership on genetic resources. Farmers could replant seed from crops under the PBRs system. However, intellectual property rights can now be claimed on the novel techniques used in genetic engineering, and also on genes and genetically modified organisms (GMOs). In terms of transgenic crops, this amounts to a wholesale privatization of major food plants. Around 90 per cent of patents involving new technologies are awarded to global corporations. Multinationals have spent the past two decades buying up independent seed companies in order to exploit patented crop seed (RAFI, 1999; RAFI, 2001c).

In contrast, public-sector plant breeding is in rapid decline and becoming increasingly marginalized. It has been estimated that the money spent by Monsanto in acquiring seed companies in 1998 alone would have funded the entire publicly owned plant-breeding programme administered by the Consultative Group on International Agricultural Research (CGIAR) – involving eighteen research stations, mainly in the developing world – at 1998 levels for more than 12 years (RAFI, 1998). Meanwhile, the original custodians of crop genetic resources, the small farmers whose knowledge and skills over many generations have produced the plants on which genetic engineers operate, have been excluded from patent considerations.

The patenting of life is putting undue strain on a patent system that was originally designed to protect inanimate inventions. Patent offices are currently struggling to process a tidal wave of highly technical applications. In Europe, for instance, patent applications rose from around 3,000 per year in the early 1970s, to over 76,000 in 1999 (RAFI, 2001c). To shoehorn many applications involving genetic modification into the system, confusion has been sown between 'invention' and an 'inventive step'. The patenting of individual genes does not even represent an 'inventive step', but merely a discovery. In many cases, the function of patented genes is not fully understood. Meanwhile, patents continue to be awarded to self-reproducing life forms that cannot be controlled in the environment. This type of patenting has been described as 'more a statement of territory than of innovation' (Doyle, 1985). As the territory in question is food production, the unprecedented control of major crops raises serious concerns about food security.

Biopiracy Developing countries have resisted the challenge of patents as best they can, but are up against powerful commercial interests backed by the World Trade Organization (WTO). Within the WTO, an agreement called TRIPs (trade-related intellectual property rights) ensures that patent rights are respected around the world. In effect, the patent system of the USA is being globalized. Countries such as India risk having their biological resources and indigenous knowledge pirated. Plants used over many generations in medicine and pest control, such as neem (*Azadirachta indica*) and turmeric (*Curcuma longa*), are becoming the intellectual property of foreign companies. RiceTec of Texas, for example, claimed Basmati rice, which has been grown for thousands of years in the foothills of the Himalayas, as a brand-name in 1997, on the basis that they had devised a new strain from existing Indian and Pakistani cultivars. Patents can be awarded to crops because a gene has been added, while centuries of plant breeding by small farmers is considered irrelevant (Shiva, 2001a).

In 1972, India stopped awarding patents on food and health products, because it considered them detrimental to the public good. The Indian government has been challenging transgenic crop patents since 1992, when it revoked a controversial patent awarded to Agracetus (Monsanto) covering all genetically engineered cotton. However, India is being forced to amend its patent laws by the WTO, which India joined in 1995. The WTO ruled that India must recognise international transgenic crop patents by 2005. Although India successfully challenged the turmeric patent, because they could show written evidence of its long history of use as a wound treatment, the odds are often with the multinational companies and their lawyers in such cases. This is partly because oral and traditional knowledge is not afforded the same respect as a recent scientific paper when it comes to awarding patents.

'Bioprospecting' or 'gene hunting' has over many decades resulted in the genetic resources of developing countries finding their way into botanical gardens and gene banks in the developed world. Multinational companies have mined these resources for genes of commercial value (Juma, 1989). The patenting of transgenic plants and other GMOs has meant that developing countries need to be even more vigilant in protecting their genetic resources. The UN's Convention on Biological Diversity has come to the aid of developing countries by defining biological resources as national property, rather than a common heritage that gene hunters can freely remove. However, existing collections in gene banks, accounting for around 90 per cent of known genetic resources for the most important food crops, may remain exempt from rulings that profits should be shared with developing countries in the centres of plant diversity. The Convention's ruling also assumes that genes can be owned as private property, rather than questioning whether genes should be patented at all. The metaphor of bioprospecting therefore hides prior use, the knowledge of indigenous communities and the right to utilize local biodiversity (Shiva, 1998).

Corporate alliances   The claim that corporations need patents to protect their investments in the face of competition is looking increasingly hollow now that mergers have left so few players in the field. These major players have a history of doing deals, involving patent licences and royalty payments, to their mutual benefit. The monetary value of alliances has typically increased, to around 20 per cent of company income, at a time of increased corporate mergers (RAFI, 2001c). The true extent of corporate oligopoly is disguised through alliances. Patents are sought not to exclude competitors, but to maximize profits.

The multinational companies that hold transgenic crop patents have already shown that they are willing to go to great lengths to protect their intellectual property rights. Due to corporate mergers, these companies are to an increasing extent the same ones that are involved in pharmaceuticals. The top ten pharmaceutical companies control more than 40 per cent of the world market. Pharmaceutical companies, including Novartis and AstraZeneca, have worked together through the courts in the past, to try to protect the markets for their patented drugs in the developing world. They have tried to prevent governments from buying cheaper generic products to treat conditions such as AIDS, diarrhoea, malaria, meningitis, and tuberculosis, which are among the most common causes of death in poor countries. If deprived of affordable drugs, millions cannot obtain the treatments they need (Boseley, 2001). In South Africa, the drug corporations backed down on one such legal action in the face of massive public protest (McGreal, 2001). These same life science corporations are likely to use legal means to protect markets for their patented transgenic crops.

Rice genome   The first plant to have its genome completely sequenced was thale cress (*Arabidopsis thaliana*) in 2000. Genes discovered in thale cress have helped to protect wheat against disease, ripen tomatoes, make citrus trees develop rapidly, and increase the yields of oilseed rape. Syngenta completed the sequencing of the rice (*Oryza sativa*) genome in the following year. Rice was the first important food crop to have its genome sequenced, and other crops will soon follow. However, the breakthrough genetic code for rice, the most important food plant in the developing world and one that has been grown and selectively bred for 5,000 years, will not be made public.

Syngenta will allow researchers limited access to data on rice's 50,000 genes, in return for a share of any commercial products arising from collaborative research. Monsanto had also completed a rough draft of the rice genome by 2001. Meanwhile, a publicly funded International Rice Genome Sequencing Project hopes to publish a more thorough sequence by 2004. Syngenta provided some initial funding for this project, while Monsanto has shared its sequencing data, but the public project has not kept pace with the commercial venture. The private sector has a couple of years to patent rice genes before the information enters the public domain. By 2001, there were already around 230 patents on rice. Patent protection will limit access and the uses to which the rice genome knowledge is put.

An appropriate technology?

Monocultures of the mind   There has arisen a fundamental conflict between global agricultural production systems and local communities in the developing world. Like the high-yielding Green Revolution crops before them, monocultures of 'first generation' transgenic crops threaten to accelerate the erosion of crop diversity, which is maintained through alternative methods of production. Minor crops and local knowledge stand to be further undermined by what Vandana Shiva, Director of the Research Foundation for Science, Technology and Ecology, has described as a 'monoculture of the mind'. This is a state of mind where only one solution can be seen, based on the commodification of agriculture. It accompanied the introduction of high-yielding Green Revolution varieties, and now applies to transgenic crops. In the context of high-yielding varieties, secondary products and the long-term sustainability of the land are undervalued, while the economic value of the primary harvest is overemphasized. High-yielding varieties, including transgenic crops, perform better because they are more responsive to inputs, in the form of irrigation, artificial fertilizers and pesticides. In traditional production systems, crop varieties may have lower yields, but they require less water resources, put much more organic material back into the soil, often produce better straw and other secondary products, and tolerate competition from 'weeds' that are often minor crops in their own right (Shiva, 1993).

There are many 'hidden costs' associated with intensive agriculture, which are not considered relevant because its commodification has put the focus exclusively on monetary costs and benefits. The hidden costs of pesticides, for example, include health problems due to pesticide poisoning, loss of beneficial insect species (e.g. predators and pollinators), the development of pest resistance, and groundwater contamination (Pearce and Tinch, 1998). If these social and environmental costs are considered, the long-term benefits of intensive agriculture do not look so superior to alternative methods of production.

The world's population is expected to increase from around six billion today to around nine billion by 2050. As many as 800 million people go hungry, and demand for food is increasing. Both Syngenta and Monsanto have promised to come up with royalty- and licence-free rice varieties for subsistence farmers. These could result in increased yields of rice, and also rice with enhanced nutritional properties. Transgenic technology, if developed in this way, could play an important role. However, it would be a mistake to concentrate on transgenic crops at the expense of alternative approaches to combating hunger and disease in the developing world.

Golden Rice Around 40 per cent of the world's population suffers from some sort of mineral or vitamin deficiency, such as lack of sufficient vitamin A, iron, iodine, selenium, or zinc. A number of transgenic crops, including vitamin and iron-enriched rice, are soon to be 'given' to poor farmers in developing countries to help combat dietary deficiencies, while at the same time being marketed as functional foods to wealthy consumers in the developed world. Golden Rice, for example, has been modified for enhanced vitamin A, via the transgenic expression of ß-carotene (Ye *et al.*, 2000; this volume Chapter 4). It has been claimed that Golden Rice will make a great contribution to solving the problems created by vitamin A-deficiency (VAD), also called xerophthalmia. Diseases associated with VAD are prevalent in over 26 countries in Asia, Africa, and Latin America. It has been estimated that more than 124 million children world-wide are deficient in vitamin A, and that a quarter of a million children go blind each year because of this nutritional deficiency. Syngenta, the company promoting Golden Rice, has claimed that in a single month 50,000 children could be cured of blindness using this technology (Brown, 2001). But how appropriate a technology is Golden Rice for solving this problem, and could it really be this effective?

Children with vitamin A deficiency tend to be undernourished, and so enhancing one vitamin in their diet will not solve their problems. VAD is usually accompanied by deficiencies in iron, iodine and other nutrients. The absorption of pro-vitamin A has been shown to depend on the overall nutritional state of the body, with the result that additional vitamins in the diet will simply be excreted in an undigested state in the undernourished. Improvements in nutrition depend on a variety of food sources, and should not be reliant on a single source. There is still much to be done to address the general problem of malnutrition. This suggests that agricultural diversity should be promoted, rather than a monoculture of transgenic rice.

There is a range of alternative approaches to combating VAD that risk being eclipsed by the promotion of Golden Rice. Projects initiated by the Food and Agriculture Organization (FAO) of the United Nations and the World Health Organization (WHO) have tackled VAD by using a combination of methods (Koechlin, 2000). General improvement of diet is a central component of these projects. Carrot, mango and dark green vegetable leaves, such as spinach, are particularly rich in pro-vitamin A, and can be highly beneficial. Women in Bengal, for example, traditionally use more than a hundred plants for green-leaf vegetables. The abundance of many of these plants, including the vitamin A-rich bathua, has been reduced by the cultivation of high-yielding varieties. The cultivation of transgenic crops could therefore reduce the abundance of plants that traditionally supply vitamin A in the diet (Shiva, 2001b). A range of other foods contain abundant vitamin A, including chicken, meat, eggs and milk. A holistic approach, involving food fortification (e.g. adding vitamins to margarine or sugar), food supplements (e.g. supplying vitamin pills twice a year), dietary education, and encouraging the cultivation and consumption of green vegetable leaves, has proved effective in FAO and WHO trial projects. In the future, transgenic crops such as Golden Rice could complement such schemes. People are unlikely through choice to want to eat just rice when a balanced diet could supply their vitamin requirements. However, in food shortages vitamin-enhanced foods could play an important role (Koechlin, 2000).

The marketing of transgenic rice as a solution to an important developing-world problem looks suspiciously like a strategy to advance the cause of transgenic crops generally, rather than delivering a workable solution on the ground in the developing world. The Rockefeller Foundation, who helped to fund the rice's development, have since distanced themselves from the project and claimed that the public relations campaign has 'gone too far' (Brown, 2001). Syngenta appears to have been exaggerating when it claimed that the rice could save the sight of over 500,000 children a year. By 2001, experimental plots of Golden Rice had levels of vitamin A that would supply 10–40 per cent of the required daily intake, based on average rice consumption for an adult (Potrykus, 2001). Scientists at the International Rice Research Institute (IRRI) in the Philippines estimated that it would be at least 2005 before the rice was ready for field trials.

There is no easy fix to problems of malnutrition caused by poverty in the developing world. Golden Rice could make a significant contribution within holistic programmes to reduce VAD diseases in the Third World, but not necessarily as the major component. It could provide an emergency foodstuff or provide a useful complement to a balanced and varied diet, while the technology could also be extended to other crops, such as cassava, sweet potato, wheat and banana. Golden Rice is far from being the 'magic bullet' solution that some have claimed. The same will be true of rice with enhanced iron, and other transgenic crops with enhanced nutrients and vitamins.

The sustainable alternative

Sustainable agriculture seeks to make the best use of available natural and social resources. Solutions are sought through an understanding of local agroecosystems; so that nutrient cycling, nitrogen-fixation and soil regeneration processes are integrated into food production systems. The use of artificial pesticides and fertilizers is minimized by replacing them with biological pest control and natural manures. The knowledge and skills of local farmers are central to this approach. Unfortunately, an assumption has arisen, backed by evidence from the Green Revolution, that sustainable agriculture is somehow old-fashioned and must necessarily be replaced by modern biotechnology-based approaches if food production is to be increased further. This assumption is now being challenged (e.g. Pretty, 1995; Pretty, 1998).

The intensification of agriculture, which increases yield per unit area of land, means that there is less pressure to convert non-agricultural land, for example in the form of biodiversity-rich forests or grasslands, to agriculture. However, the environmental costs of intensive agriculture can be great. A study of sustainable agriculture, by Jules Pretty and his colleagues at the University of Essex in England, concluded that if sustainable farming was adopted on a much larger scale it could also lead to substantial increases in food production. This would similarly take the pressure off natural ecosystems, but without causing further environmental damage.

The University of Essex study looked at 208 projects in 52 countries of Africa, Asia and Latin America which covered a combined area of about 30 million hectares, representing about 3 per cent of the agricultural land in the developing world, or an area equivalent in size to Italy. The adoption of sustainable agriculture in these projects led to yield increases of between 50 and 150 per cent (average 73 per cent), for non-irrigated crops. In 88 per cent of cases, it was reported that better use was being made of local natural resources. Improvements were also seen in education and the use of human resources, with rural livelihoods being safeguarded. In many cases, large reductions in pesticide use were reported, together with economic and environmental benefits (Pretty, 1995; Pretty, 2001). The adoption of sustainable agriculture can therefore increase yields by making the best use of natural processes, such as soil regeneration, nutrient cycling and biological control, and human resources.

Soil fertility   Sustainable agriculture places particular emphasis on maintaining soil fertility, by making improvements to organic content, physical structure, water-holding capacity and nutrient balance, and by preventing soil erosion. One sustainable method for achieving these aims is 'zero tillage', whereby ploughing is eliminated or reduced to a basic minimum. Zero tillage is being adopted to an ever greater extent in the developing world, particularly in Latin America. In Brazil, for example, the area under zero tillage or '*plantio directo*' increased from one to eleven million hectares between 1991 and 1999, while one third of Argentinian farms had turned their back on the plough by 1999. In global terms, zero tillage is also beneficial because unploughed soil holds its carbon, while ploughing releases it to the atmosphere in the form of carbon dioxide. A one-hectare field left unploughed absorbs up to a tonne of carbon every year, which is an important factor in moderating climate change (Pretty, 2001; Pearce, 2001).

Polyculture   Sustainable agriculture embraces diversity and polyculture, while transgenic crops are grown in monocultures. Intercropping, the simultaneous cultivation of two or more different crops on the same area of land, can reduce the incidence of pests and diseases, reduce weed problems, maintain soil fertility, reduce soil erosion, and produce high combined yields. In Africa around 80 per cent of cereals are intercropped, and in Latin America about 60 per cent of maize is intercropped (Vandermeer, 1989).

In a monoculture, every non-crop plant is considered a weed, but other plants can be considered beneficial in sustainable agriculture. In East Africa, thousands of farmers have taken to planting a local plant called Napier grass in their maize fields, because it attracts stem-boring insect pests away from the crop. Meanwhile, a damaging parasitic weed called *Striga* can be controlled in maize by planting another weed called *Desmodium* between the crop rows, which releases a chemical that prevents *Striga* from growing. In Kenya, these sustainable methods have proved cheaper and better at controlling pests and weeds than pesticides, and have resulted in yield increases of up to 70 per cent (Pearce, 2001).

Sustainable agriculture can therefore provide an environmentally and socially beneficial alternative to intensive agriculture. The myth that modern high-tech farming always represents the most productive method is starting to be dispelled. Biodiverse farms can be at least as productive as industrial monocultures. Sustainable methods are ideally suited to small farms and poorer farmers, who cannot afford pesticides but who are knowledgeable about local farming conditions. Real improvements could be achieved if sustainable methods could be extended, from the current 3 per cent to 30 per cent or more of the agricultural land in Africa, Asia and Latin America. However, without policy support, such methods may remain localized, and may become overshadowed by transgenic crop technology (Pretty, 2001).

The organic movement   In the developed world, intensive agriculture is increasingly contrasted with sustainable systems of production. A cultural change is under way, illustrated by the rise in numbers of consumers seeking organic and 'GM-free' foods. Organic farming is the production of food without the use of artificial fertilizers or pesticides. Organic farming is a form of sustainable agriculture, although not all sustainable agriculture is organic. The modern organic movement in Europe is thriving in the backlash against modern intensive agriculture and concerns about the environment, animal welfare, and food quality. Whilst demand for human and animal food produced from transgenic crops has plummeted in Europe, there is a rapidly growing market for organic food. In Sweden and Austria, 10–11 per cent of farms were organic by 2001, and other countries such as Denmark and Germany were pushing to catch them up. In Denmark, for example, the government gives financial encouragement to farmers to convert to organic production, while the country is profiting by exporting large amounts of organic food to countries such as Britain, where organic farming is still relatively marginalized. The level of funding for organic research in most developed countries remains only a small fraction of that available for research into transgenic crops.

Mass-produced vegetables, such as broccoli, can have far fewer minerals and vitamins than traditionally grown ones. Tomatoes bred for mechanical harvesting have around 15 per cent less vitamin C, and potatoes up to 30 per cent lower levels of riboflavin and potassium, than plants grown using non-intensive traditional methods (Doyle, 1985). Although a number of nutritional and health claims for organic food have been exaggerated, organic foods are generally of good quality and are produced by agricultural methods differing from the high-input systems typical of 'first generation' transgenic crops. Consumers are favouring the organic alternative, leaving the market for foods produced from transgenic crops in doubt.

Prospects for transgenic crops

The unprecedented uptake of transgenic crops, with a 25-fold increase in area between 1996 and 2000, reflects grower satisfaction (James, 2000). Farmers with large-scale and modernized operations, in particular, are benefiting from transgenic crops. They will continue to do so, at least in the short term. Yield increases from the 'first generation' of transgenic crops have typically been 5–20 per cent. However, these are well below those recorded during the Green Revolution, which helped to raise crop yields by up to 50 per cent. The need for such yield increases in the developed world is not always clear. Maize growers in the USA, for example, have questioned the need for the 10 per cent yield increase that transgenic crops provide, because the added production has driven down US maize prices (ACGA, 2000).

In the developing world, some yield increases with transgenic crops of over 40 per cent have been recorded, for example virus-resistant potatoes on Mexican smallholdings. In such cases, transgenic crops could make an important contribution to raising yields. In general, however, transgenic crops alone cannot solve the world's food problems. The solution must also include sustainable agriculture and crop diversification. In any case, a lack of food is rarely the underlying cause of starvation. Famines continued throughout the Green Revolution, due to poor distribution of the additional food. During African famines, affected countries continued to export cash crops to the developed world. The USA grows abundant food, but many go hungry. Food supplies would be sufficient if shared more equitably. Starvation is essentially caused by poverty, and is exacerbated by inadequate food distribution, political instability, and a complex range of social factors. If solutions are not also sought in these areas, then no level of yield increase will feed the world's hungry (Sen, 1981; Conway, 1997).

A range of environmental concerns, including genetic contamination and impacts on non-target organisms, have emerged that may slow the rate of uptake of transgenic crops. The area under transgenic canola in Canada in 2000, for example, fell in comparison to the previous year. Ultimately, questions of liability may be the most important limiting factor on the deployment of genetic engineering in agriculture. Without a 'predictive ecology' in place, insurance companies may be reluctant to issue protection against claims of damage to the environment or health (Rifkin, 1998). With the expansion of transgenic crops, transgenes are going to be found with increasing frequency in non-transgenic crops. Litigation involving genetic pollution has already started. Foods have been contaminated, while organic growers have had their livelihoods threatened. Compensation could be very expensive for the life science corporations if they are held accountable. Transgenic crops have an important role to play within agriculture, but that role may be much more limited than many originally envisaged.

The mistakes of the Green Revolution, however, need not be repeated. Transgenic crops could expand their role in the future through cross-breeding with numerous landraces, for instance, so that they are better adapted to local conditions. They could also become less reliant on high inputs of artificial fertilizers and pesticides. Modifications that make crops resistant to drought (instead of dependent on irrigation), high saline levels in soil, and other stress factors associated with marginal land would also enable subsistence farmers to reap the benefits of transgenic technology. However, this should not be done as a fix to detract from the root causes of desertification or salinification, which may include the inappropriate cultivation of irrigated crops.

Transgenic crops do not have to be incompatible with sustainable agriculture, and could be made to work alongside organic or other sustainable systems of farming. They need not be a threat to diversity. In the hands of publicly funded institutions, for example, locally adapted cultivars containing transgenes could be supplied to poorer farmers for cultivation within biodiverse systems. This marriage of transgenic and sustainable agriculture corresponds with the original vision of some of the pioneers of this technology, before it was bought wholesale by big business. Professor M.S. Swaminathan, chairman of the Swaminathan Research Foundation in Chennai (Madras), southern India, has started to develop mixed organic and transgenic farming methods, using salt-tolerant varieties of rice and other crops. A limited role for transgenic crops within a diversified 'people friendly' sustainable agriculture is the way forward, if the benefits of this technology are to be made widely available with minimal adverse social and environmental impacts (Vidal, 2001).

Genetic engineering, however, is currently a technology in corporate hands. Crop seed is controlled by a handful of multinational companies, and intellectual property rights pervade everything. Licensing agreements and Terminator technology threaten the rights of farmers to save and replant seed. Most research in this area, including that done in universities, is privately funded. The private sector has been reluctant to put in place measures to reduce the potential ecological risks associated with GMOs, because they are seen as costly and could delay the deployment of the technology. Corporate culture therefore dictates how the technology is being developed, and too little account is taken of its socio-economic impacts. The charity ActionAid identified ripening-controlled GM coffee as a typical transgenic crop in this respect. These coffee plants are being developed so that a proprietary chemical can be sprayed on to them so that they ripen simultaneously to facilitate machine harvesting. At present, coffee is hand picked and has staggered ripening times. The introduction of this transgenic coffee could therefore destroy the livelihoods of many smallholder farmers and contribute to poverty in the developing world.

For 'third way' approaches, such as the one advocated by Swaminathan, to become a reality there is a requirement for local, publicly-funded research and the active participation of communities. A new balance between the private and public sectors is needed to achieve this. Public–private partnerships would enable governments to legislate better and shape how GM technology is developed. There is a need, for example, for disease-resistance in traditional crops such as sorghum, millet and cassava. There is some evidence of this happening. Monsanto's collaboration with Kenyan researchers to develop virus-resistant sweet potatoes is one such case. However, such examples do not represent the main thrust of the present GM enterprise.

The voices of subsistence farmers and those who stand to benefit most from agricultural improvement have to be listened to, if transgenic crop technology is to make positive contributions to supplying food where it is most needed. A diversity of 'people friendly' transgenic crops should be developed, instead of a few that benefit only large industrialized operations, within a framework of environmental and community sustainability. If transgenic crops cannot be made compatible with an agriculture that seeks to reverse the trends toward environmental degradation and undermined rural societies, however, then there is a case for diverting funds to develop modern sustainable agricultural methods that can provide long-term food security.

13 The precautionary principle

The cases of 'genetic pollution' and unexpected impacts on non-target species outlined in this book suggest that more caution is required when releasing genetically modified organisms (GMOs) into the environment. Risk-management strategies involving the precautionary principle are emerging to help to achieve this. This chapter explains what the precautionary principle is and lists a number of areas where precautionary action is required. There are obstacles in the way of a precautionary approach being more widely adopted for GMO releases, however, which need to be overcome if the benefits of this technology are to be realized without unacceptable social and environmental costs.

Defining the principle

The precautionary principle states that potential environmental hazards should be dealt with even in the absence of scientific certainty. It is a principle on which to base risk-management decisions. The aim is to reduce risk to a minimum, by taking an anticipatory rather than a reactive approach. Under the principle, lack of hard proof is not an obstacle to taking precautionary action if there is suspected harm. Some scientific evidence pointing toward suspected harm is usually required before precautionary action is initiated. Precautionary action is particularly advisable if the suspected harm is irreversible, cumulative, or extending over large areas or long periods. The precautionary principle has proved particularly useful in determining action when a technology has had unexpected effects, such as when evidence first suggested that chlorofluorocarbons (CFCs) were damaging the ozone layer.

The precautionary principle was first brought into play in European environmental policy during the late 1970s, and since 1992 it has provided the basis for European environmental law. When applied to GMOs, the precautionary principle accommodates the incomplete ecological knowledge available, the time lag that can exist between evidence of harm and irrefutable proof of harm, and the possibly permanent and uncontrollable consequences of GMO release. Within the European Union (EU), the precautionary principle has been invoked through a clause (Article 16) in

Directive 90/220 on the deliberate release of GMOs into the environment, by at least six member states to ban imports of certain transgenic crop seed and GM food. The clause was designed as a short-term measure, pending arbitration, for countries to object to imports that had been approved by the EU, but which they considered unsafe. Austria and Luxembourg, for example, used the clause to ban the import and cultivation of Novartis' *Bt* maize (corn). They argued that its benefits were outweighed by potential health risks, due to the presence of antibiotic-resistance marker genes, and potential adverse effects on organic growers, due to the possible development of insect-resistance to *Bt* bioinsecticides. Although there is no scientific proof to show that antibiotic-resistance genes can be transferred from plants to bacteria in human guts, there was enough evidence to justify precautionary action.

Reasons to be cautious

This book has surveyed the ecological risks associated with a range of GMOs that have been, or may soon be, released into the environment. This section summarizes twenty areas where concern has arisen and where precautionary action is desirable.

	Risk	Action
1. Microorganisms and soil ecology	Adverse effects on growth of mycorrhizal fungi and a wide range of plant species.	More microcosm studies with microbial GMOs, under conditions that mimic real soil, to look for effects on: i) mycorrhizal fungi and root systems of non-target plant species, and ii) biogeochemical cycles. Such data to be required before approval for environmental release.

	Risk	Action
2. Crop-seed production	Contamination of conventional seed with transgenic seed.	Stricter seed segregation to protect crop integrity. Mandatory labelling of transgenic seed worldwide. Particular caution required with transgenic crops not approved for human consumption.
3. Crop-seed spillage	Establishment of wild populations of transgenic plants (e.g. oilseed rape) which could act as sources of "genetic pollution".	Improved containment methods during seed transportation and storage.
4. Herbicide-resistant crops	Adverse impacts on biodiversity due to increased spraying of herbicide.	More research needed on environmental impacts of transgenic crop cultivation. In areas where biodiversity at risk, measures should be taken to protect the environment, for example, by leaving crop margins unsprayed or banning the cultivation of certain transgenic crops.
5. Insect-resistant crops	Insect pests may rapidly acquire resistance to Bt toxins, including bioinsecticides used by organic growers.	Mandatory use of resistance management strategies, such as insect refuges. Close monitoring of resistance in field and more research on assumptions underlying refuge approach.

	Risk	Action
6. "Genetic pollution"	Gene flow can result in non-transgenic crops acquiring transgenes. Possible creation of 'superweeds' or contamination of food.	Clearer guidelines needed on which transgenic crop lines should not be grown adjacent to each other. GM-free zones for conventional crops.
7. Organic and seed crop production	"Genetic pollution", or even the perception of it, can severely affect the livelihoods of organic farmers. The integrity of seed crops could be adversely affected by "genetic pollution".	Socio-economic consequences, in addition to biological ones, should be taken into account when evaluating precautionary action. Mechanisms for banning transgenic crop cultivation in certain areas, to encourage organic production and ensure purity of seed crops, should be put into place.
8. Non-universality of risk assessment	Potential adverse environmental impacts may not be accounted for if risk assessment done only in one geographic region. Crops might have more wild relatives in some countries, for example, increasing the risk of inappropriate transgene flow.	Need for greater recognition that environmental risk varies with region. International agreements should enable countries to take unilateral precautionary action regarding GMOs and this should be afforded predominance over conflicting "free trade" agreements.

	Risk	Action
9. Field trials of transgenic crops	Unpredicted ecological impacts can occur with commercial transgenic crops.	The collection of ecological data from field trials, in addition to crop performance data, should be made a necessity before approval for commercial release of transgenic crops.
10. Genetic resources	Transgene contamination in gene banks and *in situ* conservation areas.	Moratorium on transgenic crop cultivation in their centres of origin and diversity.
11. Antibiotic-resistance marker genes	Potential spread of antibiotic-resistance from GMOs to bacteria in guts of animals or humans, leading to reduced effectiveness of antibiotics in veterinary and medicinal use.	Measures to phase out antibiotic-resistance marker genes and to develop alternatives should be speeded up. Countries that wish to invoke the precautionary principle, to ban imports of transgenic crops containing certain antibiotic-resistance marker genes, should be allowed to do so.
12. Promoters	Promoters (e.g. CaMV 35S) may lead transgenes to become active in inappropriate situations (e.g. in pollen).	More research is needed on the biosafety of transgene promoters. Promoters should be taken into greater account when assessing the risks of transgenic crops. A move towards tissue-specific transgene expression should be encouraged.

	Risk	Action
13. Virus-resistant crops	Viral transgenes could recombine with plant viruses to form novel strains of pathogenic virus.	Bans on the use of certain viral transgenes should be extended into precautionary legislation worldwide.
14. Allergens	Unexpected allergens could occur in food due to the genetic modification process or through contamination of crops with transgenes in the field (e.g. in the StarLink maize case).	A more systematic scheme for testing foods for the presence of transgene-derived proteins is required, for instance, in honey produced near transgenic crops.
15. Non-target organisms	A range of non-target organisms (e.g. natural enemies, pollinators) could be affected by transgenic crop cultivation. This type of environmental impact is only being picked up after crops are grown commercially (e.g. in the case of the monarch butterfly and *Bt* maize).	Tri-trophic interactions (e.g. plant–herbivore–predator) need to be taken into account during the evaluation of transgenic crops. The registration process for commercial release needs to include a much fuller consideration of ecological interactions.
16. Invasiveness	Genetic modification may lead to increased invasiveness, to the detriment of native species.	Ecological models that predict potential invasiveness of GMOs should be more widely applied prior to commercial release into the environment.

	Risk	Action
17. Trees	Pollen containing transgenes could disperse over long-distances. Transgenes could become established in native trees with adverse ecological impacts.	Until more research is conducted, for example, on gene flows between transgenic and unmodified trees, moratoriums on commercial releases of transgenic trees should be put in place.
18. Fish	Large numbers of transgenic fish are likely to escape from fish farms, with adverse ecological impacts.	Containment of transgenic fish needs to be considerably improved compared to most existing aquaculture operations. Modifications for sterility should be encouraged where appropriate. Moratoriums on the release of transgenic fish within sea cages are required.
19. Mammals	Unplanned release of transgenic animals may have adverse ecological impacts.	Procedures need to be adopted to deal with unplanned releases of transgenic animals.
20. Intellectual property	The desire to safeguard intellectual property rights may lead multinational corporations to introduce measures that are detrimental to poorer farmers, such as Terminator technology.	Socio-economic and crop conservation factors need to be considered prior to commercial approval. Ban on routine use of Terminator technology.

The Biosafety Protocol

All around the world, GMOs that have no respect for national boundaries are being released into the environment. It is imperative, therefore, that international agreements on the safety of GMOs are put in place. The Biosafety Protocol, which arose from the second United Nations Biodiversity Convention (Jakarta, Indonesia, 1995), was an attempt by the international community to control the risks associated with the release of GMOs into the environment. However, representatives from 135 countries failed to reach an agreement at the subsequent meeting (Cartagena, Colombia, 1999) partly because governments, industry and nongovernmental organizations strongly disagreed about the ecological risks posed by transgenic crops. The major grain-exporting countries, the so-called 'Miami bloc' (Canada, Australia, Argentina, Chile, Uruguay and the USA) successfully stalled the Protocol. The USA was a 'non-party' to the Biosafety Protocol, having never signed it, but it carried immense influence through its lobbying activities (Greenpeace, 1999).

The precautionary principle was eventually incorporated into the Biosafety Protocol agreed at the Biodiversity Convention in Montreal, Canada, in 2000. This enabled nation states to ban the import of GMOs if their safety could not be conclusively established and hazards were suspected. The Protocol also enabled countries to take into account potential 'socio-economic factors' arising from the impacts of GMOs. However, the 'Miami bloc' succeeded in ensuring that this provision applied only to GMOs, such as transgenic crop seed and modified microorganisms, and not to processed products such as GM foods.

The science conducted in one country may not be applicable to all countries, because environmental conditions and circumstances differ. Approvals for release of GMOs into the environment (or marketing consents) granted in one country, therefore, may not be appropriate to another country, because different risk factors might be present. Approval for release in the USA does not mean that a transgenic crop can be cultivated globally without ecological risk. Around a third of the major commodity crops in the USA (e.g. soyabeans, maize, cotton) are transgenic and relatively few problems have been reported. However, in the USA these transgenic crops are not native and have few wild relatives. Developing countries, on the other hand, may be located in the centres of genetic diversity for particular crops, or have many landraces and wild relatives with which a transgenic crop could hybridize. Transgenes could in such cases quickly establish outside cultivated areas with potentially serious consequences for the conservation of genetic resources. Nations can therefore use the precautionary principle to help them to take uni-lateral decisions regarding GMOs, depending on their own particular situation. It has proved valuable as a basis for regulation in many developing countries, such as India, Kenya and South Africa. However, many poorer countries regulate imports based on what the USA considers safe, because they cannot afford to do otherwise. Meanwhile, food aid containing GM soya and maize can spread transgenes.

Too few precautions have been taken to date when assessing the risks of releasing GMOs into the environment. The ecological risks in some cases call into question whether the GMO should have been released into the environment in the first place. Other GMOs may have been released prematurely, before appropriate management strategies had been fully formulated. The emerging evidence of ecological risk justifies the moratoriums on commercial releases of transgenic crops put in place by certain countries, while further research is carried out on their environmental impacts. There is a need for co-ordinated research in this area, preferably with independent funding, but ecological studies take time and slow down approvals for GMO releases. Commercial interests, meanwhile, understandably want to push ahead to realize profits on their substantial investments.

Part of the problem to date with the precautionary principle has been its vague definition. It has too often been dismissed as incoherent, impractical, or too expensive to apply. It has also been open to multiple interpretations. Corporate responses to the precautionary principle stem from environmental management plans that tend to view environmental protection as an add-on to a 'business as usual' approach, rather than an agenda for changing how business is done. Corporations have in this way remodelled the environmental agenda to fit in with how they operate. World business organizations, such as the International Chamber of Commerce, in their Charter for Sustainable Development have adopted a weak definition of the precautionary principle (Welford, 1997). Too strong a formulation of the precautionary principle is undesirable, as this would prevent any technological advance. Industry is right to stress the need for some scientific evidence of potential harm before the principle is invoked. However, many weak definitions of the precautionary principle require what almost amounts to scientific proof, which is equivalent to having no precautionary principle at all.

There is a need for a standard definition of the precautionary principle so that it can be applied consistently on an international basis. To this end, the EU set out guidelines in February 2000 for applying it. These stressed the need for reliable scientific data and logical reasoning before triggering the use of the principle, the necessity for identifying potentially hazardous effects, and the need for both an appropriate cost-benefit analysis and an assessment of the degree of scientific uncertainty. Precautionary action was seen as a provisional measure pending the availability of more scientific data (Foster *et al.*, 2000). The precautionary principle is therefore based on sound science, is not emotional or irrational, and will not place undue financial burdens on nations or corporations if applied judiciously.

There is strong resistance to taking a more cautious approach to the release of GMOs into the environment, however, orchestrated by some governments, corporations, and international business organizations. Ranged against the Biosafety Protocol, for example, are free trade agreements enforced by the World Trade Organization (WTO). In the name of 'free trade', bans on imports can be challenged, even though those bans have resulted from a precautionary approach being taken to new technology. This is evident in conflicts between the USA and Europe over beef from animals fed with growth hormones, and transgenic crops. Although many see the precautionary principle as an important brake that can be applied to the biotechnology juggernaut, powerful industry interests and other critics of the principle see it as a veiled form of protectionism and an irrational limitation to the spread of biotechnology.

Few US laws quote the precautionary principle. Industry self-regulation has been seen as adequate, while there has been an insistence that genetic engineering is merely an extension of traditional plant breeding and requires no additional safety measures. Existing regulations covering plant pests, pesticides, and food and drugs, have therefore been used to regulate GMOs. The US Department of Agriculture (USDA) has approved nearly 30,000 field trials involving transgenic crops, but has not taken into account unique biosafety risks, such as those involving gene flow or the presence of antibiotic-resistance marker genes. The precautionary principle has played no role in what has become little more than a rubber-stamping exercise. The gulf between the way GMOs are regulated in the USA and other 'Miami bloc' countries compared to Europe and elsewhere may widen, with the precautionary principle being a source of friction for some time to come.

Landscapes and genescapes

In this book, my aim has been to summarize recent research on the ecology of genetic engineering. The debate can be moved forward only if emerging knowledge is assimilated and understood. The emphasis in this book has therefore necessarily been on the potential risks of the technology. It is illogical to denounce entire areas of biotechnology without assessing the potential costs and benefits of each application. The development of transgenic crops and other GMOs should continue, as there are many benefits that could result from their release into the environment, particularly in terms of productivity and enhanced nutritional or medicinal properties of food. To date however, many of these benefits remain largely speculative, while a range of potential environmental and social costs are not being sufficiently addressed.

Genetic engineering involves inserting genes into genomes. It is an imprecise technique, in that transgenes insert at random into genomes and the success rate is often low (hence the need for selectable marker genes). A range of mechanisms at the molecular level means that there is a degree of unpredictability about the process. In the environment, GMOs may be uncontrollable once released, and any ecological effects may be irreversible. Running through this book has been the idea that genes 'escape' from organisms, and in ways that are often hard to predict or to control. Microorganisms may be effectively impossible to retrieve, for instance, and can freely exchange genes with other microorganisms. Transgenes from transgenic crops have established in other cultivars, beyond the control of seed companies and growers. Pollen from transgenic trees could fertilize native trees over long-distances. Transgenic fish might readily escape from sea cages and interbreed with wild populations. The uncertainties about the long-term ecological risks of GMOs suggest that decisions to approve their release into the environment should err on the side of caution.

Ecologists adopt new analytic tools and methods as they are developed in other disciplines. One such case involves the merging of traditional ecology with the latest genome technology in a new field called ecological genomics. The aim of researchers in this new branch of ecology is to identify the function of an organism's genes in ecological terms. The activity patterns of identified genes can reveal an organism's metabolic and nutritional needs, for example, and supply clues about its role within an ecosystem. Eventually, it might be possible to analyse ecological systems at the genomic level. An early application of this approach involved the analysis of bacterial DNA from the ocean off California. A gene sequence was compiled from the DNA obtained, which revealed the presence of genes that coded for rhodopsin, a light-absorbing pigment involved in photosynthesis. The data led to a more accurate model of part of the carbon cycle at the ocean surface (Leslie, 2001). As the techniques of genetic analysis get ever more sophisticated, ecologists can start to study ecosystems in terms of the genes present in organisms, populations and communities.

Mankind has now modified the planet to a remarkable extent. A view from space shows that 24 per cent of the Earth's available land surface has been cultivated or paved (AAAS, 2001). Human beings are an integral part of the Earth's ecology and few ecological systems remain uninfluenced by human activity. Many species of microorganism, plant, and animal have been selectively bred throughout human history. Numerous species have been transported around the world through deliberate or accidental human intervention, causing invasions that have had major ecological impacts. Agricultural expansion has proceeding with the ultimate aim of feeding just one species, while other species have had to adapt to changing habitats. In some cases, species have become better adapted by acquiring genes from cultivated plants and domesticated animals. Other species have adapted to urban landscapes. These adaptations all involve genetic change. Many other species have become extinct. With the release of numerous self-reproducing GMOs into the world's ecosystems, human activity can be expected to have an increasing influence on both landscapes and ecological genescapes.

Genetic engineering therefore represents the latest means by which humans can modify the genetic make-up of other organisms. It is a particularly powerful means because for the first time genes from one species can be integrated into the genomes of completely unrelated species. The experience with GMOs to date suggests that some transgenes may soon become ubiquitous in the environment. Will this have any adverse long-term ecological effects? Studies of gene ecology may soon be able to answer such questions, but there is currently insufficient knowledge available to do so.

Ultimately, ecosystems are complex systems, and within them everything is interconnected. Changing the genetics of an organism can potentially affect other species via a series of ecological relationships. This is nothing new, because mankind has been genetically adapting other species since the first days of agriculture. However, as we enter an age of unprecedented opportunity for modifying the inherited characteristics of organisms, the possible ecological consequences of doing so should constantly be borne in mind.

Bibliography

[Numerals in square brackets indicate chapter in which reference is cited]

AAAS (American Association for the Advancement of Science) (2001) *AAAS Atlas of Population and Environment*, Berkeley: University of California Press. [13]

Abdel-Mallek, A.Y., M. Abdel-Kader and A. Shonkeir (1994) 'Effect of glyphosate on fungal populations, respiration and the decay of some organic matters in Egyptian soil', *Microbiology Research*, 149: 69–73. [4]

ACGA (American Corn Growers Association) (2000) 'Corn growers assert that increased corn yields is the wrong reason to plant genetically modified corn', <http://www.biotech-info.net/> [12]

ACNFP (Advisory Committee on Novel Foods and Processes) (1994) *Report on the Use of Antibiotic Resistance Markers in Genetically Modified Food Organisms*, London: MAFF. [8]

ActionAid (2001) 'Wake up and smell the GM coffee', <http://www.actionaid.org.uk/> [8]

Ainsworth, C. (2001) 'Grow fast, die young', *New Scientist*, 17 February, p. 14. [10]

Ali, A., E.P. Fuerst, C.J. Arntzen and V. Souzha-Machado (1986) 'Stability of triazine resistance in Rutabaga backcross generations', *Plant Physiology*, 80: 511–14. [8]

Al-Kaff, N.S., S.N. Covey, M.M. Kreike, A.M. Page, R. Pinder and P.J. Dale (1998) 'Transcriptional and posttranscriptional plant gene silencing in response to a pathogen', *Science*, 279: 2113–15. [11]

Allison, R.F., A.E. Greene and W.L. Schneider (1997) 'Significance of RNA recombination in capsid protein-mediated virus-resistant transgenic plants', in M. Tepfer and E. Balzács (eds.), *Virus-resistant Transgenic Plants: Potential for Ecological Impacts*, Versailles & Heidelberg: INRA & Springer-Verlag, pp. 40–4. [4]

Anderson, C. (1993) 'Flood uproots transgenic crop', *Science*, 261: 1271. [11]

Anon. (1997) (Monsanto advertisement), *Farm Journal*, November, B–25. [8]

Anon. (1998) 'Promiscuous pollination', *Nature Biotechnology*, 16: 805. [6]

Anon. (2000a) 'Weed shift worries', *Progressive Farmer*, 12 June, <http://www.biotech-info.net/> [4]

Anon. (2000b) 'Resistance is useless', *New Scientist*, 19 February, p. 21. [6]

Anon. (2001a) 'Nowhere to hide', *New Scientist*, 10 February, p. 15. [4]

Anon. (2001b) 'The conflict' and 'Canadian farmer Percy Schmeiser loses canola gene case', <http://www.percyschmeiser.com/> [6]

Arthur, C. (1999) 'Farmer kills GM crops after threats', *Independent*, 8 June. [11]

Atlas, R.M. (1995) 'Bioremediation', *Chemical and Engineering News*, 73 (14): 32–42. [3]

Austin, A.P., G.E. Harris and W.P. Lucey (1991) 'Impact of an organophosphate herbicide (glyphosate) on periphyton communities developed in experimental streams', *Bulletin of Environmental Contamination and Toxicology*, 47: 29–35. [4]

Austin, H.K., P.G. Hartel and D.C. Coleman (1990) 'Effects of genetically-altered *Pseudomonas solanacearum* on predatory protozoa', *Soil Biology and Biochemistry*, 22: 115–17. [3]

Avery, M. and D. Gibbons (1999) 'Test of the revolution', *Guardian (Society)*, 18 August. [7]

Bailey, M.J., N. Kobayashi, A.K. Lilley, B.J. Powell and J.P. Thompson (1994) 'Potential for gene transfer in the phytosphere: isolation and characterization of naturally occurring plasmids', in M.J. Bazin and J.M. Lynch (eds.), *Environmental Gene Release: Models, Experiments and Risk Assessment*, London: Chapman & Hall, pp. 77–98. [3]

Barbosa, P. and D.K. Letourneau (1988) 'Conceptual framework of three-trophic-level interactions', in L.P. Barbosa and D.K. Letourneau (eds.), *Novel Aspects of Insect–Plant Interactions*, New York: John Wiley & Sons, pp. 1–90. [7]

Barton, K.A. and M.J. Miller (1993) 'Production of *Bacillus thuringiensis* insecticidal proteins in plants', in S.D. Kung and R. Wu (eds.), *Transgenic Plants Vol. 1: Engineering and Utilization*, New York: Academic Press, pp. 297–315. [4]

Bartsch, D., M. Schmidt, M. Pohl-Orf, C. Haag and I. Schuphan (1996) 'Competitiveness of transgenic sugar beet resistant to beet necrotic yellow vein virus and potential impact on wild beet populations', *Molecular Ecology*, 5: 199–205. [5]

Bartsch, D., H. Sukopp and U. Sukopp (1993) 'Introduction of plants with special regard to cultigens running wild', in K. Wöhrmann and J. Tomiuk (eds.), *Transgenic Organisms: Risk Assessment of Deliberate Release*, Basel: Birkhäuser Verlag, pp. 135–51. [5, 12]

Begon, M., J.L. Harper and C.R. Townsend (1990) *Ecology: Individuals, Populations and Communities*, 2nd Edition, Oxford: Blackwell Scientific. [2]

Bergelson, J. (1994) 'Changes in fecundity do not predict invasiveness', *Ecology*, 75: 249–52. [6]

Bergelson, J. and C.B. Purrington (1996) 'Surveying the costs of resistance in plants', *American Naturalist*, 148: 536–58. [6]

Bergelson, J., C.B. Purrington, C.J. Palma and J.-C. López-Gutiérrez (1996) 'Costs of resistance: a test using transgenic *Arabidopsis thaliana*', *Proceedings of the Royal Society of London, Series B*, 263: 1659–63. [6]

Bergelson, J., C.B. Purrington and G. Wichmann (1998) 'Promiscuity in transgenic plants', *Nature*, 395: 25. [6]

Beversdorf, W.D., J. Weiss-Lerman, L.R. Erickson and V. Souzha-Machado (1990) 'Transfer of cytoplasmically inherited triazine resistance from bird's rape to cultivated oilseed rape (*Brassica campestris* and *B. napus*)', *Canadian Journal of Genetics and Cytology*, 22: 167–72. [4,6]

Bioengineering Action Network (2000) 'Genetically altered mutant trees on the horizon', <http://www.forests.org/> [9]

Birch, A.N.E., I.E. Geoghegan, M.E.N. Majerus, J.W. McNicol, C.A. Hackett, A.M.R. Gatehouse and J.A. Gatehouse (1999) 'Tri-trophic interactions involving pest aphids, predatory 2-spot ladybirds and transgenic potatoes expressing snowdrop lectin for aphid resistance', *Molecular Breeding*, 5: 75–83. [7]

Bishop, D.H.L., P.F. Entwistle, I.R. Cameron, C.J. Allen and R.D. Possee (1988) 'Field trials of genetically-engineered baculovirus insecticides', in M. Sussman, C.H. Collins, F.A. Skinner and D.E. Stewart-Tull (eds.), *The Release of Genetically-Engineered Micro-organisms*, London: Academic Press, pp. 143–79. [3]

Black, K.D. (1998) 'The environmental interactions associated with fish culture', in K.D. Black and A.D. Pickering (eds.), *Biology of Farmed Fish*, Sheffield: Sheffield Academic Press, pp. 284–326. [10]

Bonner, J. (1997) 'Hooked on drugs', *New Scientist*, 18 January, pp. 24–7. [8]

Boseley, S. (2001) 'At the mercy of drug giants: Millions struggle with disease as pharmaceutical firms go to court to protect profits', *Guardian*, 12 March. [12]

Bosworth, A.H., M.K. Williams, K.A. Albrecht, R. Kwiatkowski, J. Beynon, T.R. Hankinson, C.W. Ronson, F. Cannon, T.J. Wacek and E.W. Triplett (1994) 'Alfalfa yield responses to inoculation with recombinant strains of *Rhizobium meliloti* with an extra copy of *dct*ABD and/or modified *nif*A expression', *Applied and Environmental Microbiology*, 60: 3815–32. [3]

Boudry, P., M. Mörchen, P. Saumitou-Laprade, Ph. Vernet and H. Van Dijk (1993) 'The origin and evolution of weed beets: consequences for the breeding and release of herbicide-resistant transgenic sugar beets', *Theoretical and Applied Genetics*, 87: 471–8. [5]

Boyce, N. (2000a) 'Taco trouble', *New Scientist*, 7 October, p. 6. [6]

Boyce, N. (2000b) 'A breed apart', *New Scientist*, 21 October, p. 6. [6, 8]

Brasher, P. (2001) 'Prepare for the '"Terminator": first release of biotech insect planned for summer', Associated Press, <http://www.biotech-info.net/> [11]

Brem, G. and M. Müller (1994) 'Transgenic animals', in N. Maclean (ed.), *Animals with Novel Genes*, Cambridge: Cambridge University Press, pp. 179–253. [11]

Brewer, R. (1994) *The Science of Ecology*, 2nd Edition, Fort Worth, USA: Saunders College Publishing, quote p. 253. [2]

Brittenden, W. (1998) 'Terminator seeds threaten barren future for farmers', *Independent on Sunday*, 22 March. [8]

Brooks, L. and P. Brown (1999) 'Felled in the name of natural justice', *Guardian*, 13 July. [9]

Brookes, M. (1998) 'Running wild. Some plants just can't keep their genes to themselves. But does it really matter?', *New Scientist*, 31 October, pp. 38–41. [6]

Brown, C. and G. Lean (2001) 'Trials of GM maize threaten unique organic crop centre', *Independent on Sunday*, 6 May. [6]

Brown, P. (1999) 'Forests in danger from GM trees says WWF', *Guardian*, 10 November. [9]

Brown, P. (2000a) 'More fish than flesh in future diet', *Guardian*, 4 October. [10]

Brown, P. (2000b) 'Monster salmon scare for fish farmers', *Guardian*, 12 April. [10]

Brown, P. (2001) 'GM rice promoters "have gone too far"', *Guardian*, 10 February. [12]

Brush, B. (2000) *Genes in the Field: On-farm Conservation of Crop Diversity*, Boca Raton, Florida: Lewis Publishers. [12]

Bryant, J. and S. Leather (1992) 'Removal of selectable marker genes from transgenic plants: needless sophistication or social necessity?', *Trends in Biotechnology*, 10: 274–5. [8]

Buhl, K. and N.L. Faerber (1989) 'Acute toxicity of selected herbicides and surfactants to larvae of the midge *Chironomus riparius*', *Archives of Environmental Contamination and Toxicology*, 18: 530–6. [4]

Bunting, M. (2000) 'Confronting the perils of global warming in a vanishing landscape', *Guardian*, 14 November. [9]

Caddick, M.X., A.J. Greenland, I. Jepson, K.-P. Krause, N. Qu, K.V. Riddell, M.G. Salter, W. Schuch, U. Sonnewald and A.B. Tomsett (1998) 'An ethanol inducible gene switch for plants used to manipulate carbon metabolism', *Nature Biotechnology*, 16: 177–84. [8]

Campbell, L.H. and A.S. Cook (eds.) (1995) *The Indirect Effects of Pesticides on Birds*, Peterborough: JNCC. [7]

Capra, F. (1996) *The Web of Life: A New Synthesis of Mind and Matter*, London: HarperCollins, p. 7. [2]

Chakrabarty, A.M. (1981) Patent No. 4 259 444, Washington DC: US Patent Office. [3]

Chakravarty, P. and L. Chatarpaul (1990) 'Non-target effect of herbicides: I. Effect of glyphosate and hexazinone on soil microbial activity, microbial population, and *in-vitro* growth of ectomycorrhizal fungi', *Pesticide Science*, 28: 233–41. [4]

Champolivier, J., J. Gasquez, A. Mességuer and M. Richard-Molard (1999) 'Management of transgenic crops within the cropping system', in P.J.W. Lutman (ed.), *Gene Flow and Agriculture: Relevance for Transgenic Crops*, Farnham: British Crop Protection Council, pp. 233–40. [11]

Charron, B. (1995) 'Escape to sterility for designer fish', *New Scientist*, 27 May, p. 22. [10]

Chatakondi, N., R.T. Lovell, P.L. Duncan, M. Hayat, T.T. Chen, D.A. Powers, J.D. Weete, K. Cummins and R.A. Dunham (1995) 'Body composition of transgenic common carp,

Cyprinus carpio, containing rainbow trout growth hormone gene', *Aquaculture*, 138: 99–109. [10]

Chee-Sanford, J.C., R.I. Aminov, I.J. Krapac, N. Garrigues-Jeanjean and R.I. Mackie (2001) 'Occurrence and diversity of tetracycline resistance genes in lagoons and groundwater underlying two swine production facilities', *Applied and Environmental Microbiology*, 67: 1494-502. [8]

Chèvre, A.M., F. Eber, A. Baranger and M. Renard (1997) 'Gene flow from transgenic crops', *Nature*, 389: 924. [6]

Chiura, H.X. (1997) 'Generalized gene transfer by virus-like particles from marine bacteria', *Aquatic Microbial Ecology*, 13: 75–83. [3]

Clements, F.E. (1916) 'Plant succession', *Carnegie Institute Washington Publication*, No. 290. [2]

Coghlan, A. (1996) 'Aphids give snowdrops the cold shoulder', *New Scientist*, 6 January, p. 18. [7]

Coghlan, A. (1999) 'Splitting headache: Monsanto's modified soya beans are cracking up in the heat', *New Scientist*, 20 November, p. 25. [4]

Coghlan, A. (2000a) 'Killer tomatoes', *New Scientist*, 23 September, p.9. [4]

Coghlan, A. (2000b) 'The next revolution', *New Scientist*, 28 October, p. 5. [8]

Coghlan, A., D. MacKenzie and D. Concar (1999) 'It's that man again', *New Scientist*, 16 October, p. 6. [8]

Cohen, M.B., M.T. Jackson, B.R. Lu, S.R. Morin, A.M. Mortimer, J.L. Pham and L.J. Wade (1999) 'Predicting the environmental impact of transgene outcrossing to wild and weedy rices in Asia', in P.J.W. Lutman (ed.), *Gene Flow and Agriculture: Relevance for Transgenic Crops*, Farnham: British Crop Protection Council, pp. 151–7. [6]

Colborn, T., D. Dumanoski and J.P. Myers (1996) *Our Stolen Future: Are We Threatening Our Fertility, Intelligence and Survival?*, London: Abacus, pp. 97–109. [3]

Collinvaux, P. (1993) *Ecology 2*, New York: John Wiley & Sons. [2]

Conway, G. (1997) *The Doubly Green Revolution: Food For All in the 21st Century*, Harmondsworth: Penguin, pp. 1–14, 287. [12]

Cory, J.S., M.L. Hirst, T. Williams, R.S. Hails, D. Goulson, B.M. Green, T.M Carty, R.D. Possee, P.J. Cayley and D.H.L. Bishop (1994) 'Field trial of a genetically improved baculovirus insecticide', *Nature*, 370: 138–40. [3]

Cox, C. (1995a) 'Glyphosate, Part 1: Toxicity', *Journal of Pesticide Reform*, 15 (3): 14-20. [4]

Cox, C. (1995b) 'Glyphosate, Part 2: Human exposure and ecological effects', *Journal of Pesticide Reform*, 15 (4): 21–7. [4]

Cox, C. (1996) 'Glufosinate: herbicide fact sheet', *Journal of Pesticide Reform*, 16: 15–19. [4]

Crabb, C. (1997) 'Sting in the tale for bees', *New Scientist*, 16 August, p. 13. [7]

Crawley, M.J. (1987) 'The population biology of invaders', *Philosophical Transactions of the Royal Society B*, 314: 711–31. [5]

Crawley, M.J. (1996) 'The day of the triffids', *New Scientist*, 6 July, pp. 40–1. [5]

Crawley, M.J. (1999) 'Bollworms, genes and ecologists', *Nature*, 400: 501–2. [5]

Crawley, M.J. and S. Brown (1995) 'Seed limitation and the dynamics of feral oilseed rape on the M25 motorway', *Proceedings of the Royal Society of London, Series B*, 259: 49–54. [5]

Crawley, M.J., R.S. Hails, M. Rees, D. Kohn and J. Buxton (1993) 'Ecology of transgenic oilseed rape in natural habitats', *Nature*, 363: 620–3. [5]

Crawley, M.J., S.L. Brown, R.S. Hails, D.D. Kohn and M. Rees (2001) 'Transgenic crops in natural habitats', *Nature*, 409: 682–3. [5]

Crecchio, C. and G. Stotzky (1998) 'Insecticidal activity and biodegradation of the toxin from *Bacillus thuringiensis* subsp. *kurstaki* bound to humic acids from soils', *Soil Biology and Biochemistry*, 30 (4): 463–70. [7]

Cresswell, J.E. (1994) 'A method of quantifying the gene flow that results from a single bumble bee visit using transgenic oilseed rape, *Brassica napus* L. cv. Westar', *Transgenic Research*, 3: 134–7. [6]

Crouch, M.L. (1998) 'How the Terminator terminates', Washington State: The Edmonds Institute, <http://www.bio.indiana.edu/people/terminator.html> [8]

Cummins, J.E. (1994) 'Dangers inherent in the process itself: The use of cauliflower mosaic virus (CaMV) 35S promoter in Calgene's Flavr Savr tomato creates hazard', London, Ontario: University of Ontario. [8]

Dado, R.G. and D.A. Lightfoot (1997) 'Nutritional quality of corn with high nitrogen', <http://www.ilcorn.org/Reports/97011603siu.htm> [4]

Daniell, H. (1997) 'Transformation and foreign gene expression in plants mediated by micro-projectile bombardment', *Methods in Molecular Biology*, 62: 453–88. [8]

Daniell, H. (1999) 'The next generation of genetically engineered crops for herbicide and insect resistance: containment of gene pollution and resistant insects', *AgBiotechNet*, Vol. 1, August, ABN 024, <http://www.agbios.com/articles/abn024/daniell.htm> [6]

Daniell, H., R. Datta, S. Varma, S. Gray and S.B. Lee (1998) 'Containment of herbicide resistance through genetic engineering of the chloroplast genome', *Nature Biotechnology*, 16: 345–52. [8]

Daniell, H. and S. Varma (1998) 'Chloroplast-transgenic plants: panacea—no! Gene containment—yes!', *Nature Biotechnology*, 16: 602. [8]

Darmency, H. (1994) 'Genetics of herbicide resistance in weeds and crops', in S.B. Fowler and J.A.M. Holtrum (eds.), *Herbicide Resistance in Plants: Biology and Biochemistry*, Boca Raton, Florida: CRC Press, pp. 263–97. [8]

Davis, B.D. (1980) 'Gene transfer in bacteria', in B.D. Davis, R. Dulbecco, H.N. Eisen and H.S. Ginsberg (eds.), *Microbiology, Including Immunology and Molecular Genetics*, Philadelphia: Harper and Row, pp. 137–52. [3]

De Block, M., J. Botterman, M. Vandewiele, J. Dockx, C. Theon, V. Gossel, N.R. Movva, C. Thompson, M. Van Montagu and J. Leemans (1987) 'Engineering herbicide resistance in plants by expression of a detoxifying enzyme', *EMBO Journal*, 6: 2513–18. [4]

DETR (Department of the Environment, Transport and the Regions, UK) (1999a) 'Voluntary agreement on GM crops extended', News Release, No. 1057, 5 November. [11]

DETR (1999b) 'Scientific Steering Committee reports on this year's GM farm-scale trials', News Release, No. 1089, 11 November. [11]

DETR (1999c) 'GM crop farm-scale evaluations of GM herbicide tolerant oil-seed rape and maize on farmland wildlife', Fact Sheet, 13 August. [11]

DETR (2000) 'Guidance on best practice in the design of genetically modified crops', Guidance Note, ACRE Best Practice Subgroup, <http://www.environment.detr.gov.uk/acre/> [8,11]

Devlin, B. and N.C. Ellstrand (1990) 'The development and application of a refined method of estimating gene flow from angiosperm paternity analysis', *Evolution*, 44: 248–59. [6]

Devlin, R.H. and E.M. Donaldson (1992) 'Containment of genetically altered fish with emphasis on salmonids', in C.L. Hew and G.L. Fletcher (eds.), *Transgenic Fish*, River Edge, NJ: World Scientific, pp. 229–65. [10]

Devlin, R.H., T. Y. Yesaki, C.A. Biagi, E.M. Donaldson, P. Swanson and W. Chan (1994) 'Extraordinary salmon growth', *Nature*, 371: 209–10. [10]

Devlin, R.H., C.A Biagi, T.Y. Yesaki, D.E. Smailus and J.C. Byatt (2001) 'Growth of domesticated transgenic fish', *Nature*, 409: 781–2. [10]

DiFazio, S.P., S. Leonardi, S. Cherry and S.H. Strauss (1999) 'Assessing potential risks of transgene escape from fiber plantations', in P.J.W. Lutman (ed.), *Gene Flow and Agriculture:*

Relevance for Transgenic Crops, Farnham: British Crop Protection Council, pp. 171–6. [9]

Dobson, A.P. and R.M. May (1986) 'Patterns of invasions by pathogens and parasites', in H.A. Mooney and J.A. Drake (eds.), *Ecology of Biological Invasions of North America and Hawaii*, Berlin: Springer-Verlag, pp. 58–77. [5]

Dogan, E.B., R.E. Berry, G.L. Reed and P.A. Rossignol (1996) 'Biological parameters of convergent ladybeetle (Coleoptera: Coccinellidae) feeding on aphids (Homoptera: Aphididae) on transgenic potato', *Journal of Economic Entomology*, 89: 1105–8. [7]

Dorey, E. (2001) 'EU seeds production proposals', *Nature Biotechnology*, 19: 192. [6]

Dow, B.D. and M.V. Ashley (1998) 'High levels of gene flow in bur oak revealed by paternity analysis using microsatellites', *Journal of Heredity*, 89: 62–70. [9]

Downey, R.K. (1999) 'Gene flow and rape – the Canadian experience', in P.J.W. Lutman (ed.), *Gene Flow and Agriculture: Relevance for Transgenic Crops*, Farnham: British Crop Protection Council, pp. 109–16. [6]

Doyle, J. (1985) *Altered Harvests: Agriculture, Genetics and the Fate of the World's Food Supply*, New York: Viking. [12]

Doyle, J.D., K.A. Short, G. Stotzky, R.J. King, R.J. Seidler and R.H. Olsen (1991) 'Ecologically significant effects of *Pseudomonas putida* PP0301 (pRO103), genetically engineered to degrade 2,4-dichlorophenoxyacetate, on microbial populations and processes in soil', *Canadian Journal of Microbiology*, 37: 682–91. [3]

Doyle, J.D., G. Stotzky, G. McClung and C.W. Hendricks (1995) 'Effects of genetically engineered microorganisms on microbial populations and processes in natural habitats', *Advances in Applied Microbiology*, 40: 237–87. [3]

Du, S.J., Z. Gong, G.L. Fletcher, M.A. Shears, M.J. King, D.R. Idler and C.L. Hew (1992) 'Growth enhancement in transgenic Atlantic salmon by the use of "all fish" chimeric growth hormone construct', *BioTechnology*, 10: 176–81. [10]

Dunham, R.A., A.C. Ramboux, P.L. Duncan and M. Hayat (1992) 'Transfer, expression and inheritance of salmonid growth hormone in channel catfish *Ictalurus punctatus*, and effects on performance traits', *Molecular Marine Biology and Biotechnology*, 1: 380–9. [10]

Eady, C., D. Twell and K. Lindsey (1995) 'Pollen viability and transgene expression following storage in honey', *Transgenic Research*, 4: 226–31. [7]

Edwards, R. (1998) 'Devilish seed', *New Scientist*, 10 October, p. 21. [8]

Edwards, R. (2000) 'Battlefield', *New Scientist*, 9 September, p. 5. [7]

Edwards, C.A. and P.J. Bohlen (1996) *The Biology and Ecology of Earthworms*, 3rd Edition, London: Chapman & Hall, pp. 300–5. [4]

Ellstrand, N.C., B. Devlin and D.L. Marshall (1989) 'Gene flow by pollen into small populations: data from experimental and natural stands of wild radish', *Proceedings of the National Academy of Sciences, USA*, 86: 9044–7. [6]

Elton, C.S. (1927) *Animal Ecology*, London: Sidgwick and Jackson. [2]

Elton, C.S. (1958) *The Ecology of Invasions by Animals and Plants*, London: Methuen. [5]

Emberlin, J., B. Adams-Groom and J. Tidmarch (1999) *A Report on the Dispersal of Maize Pollen*, Soil Association (UK), <http://www.soilassociation.org/> [6,9]

EPA (Environmental Protection Agency) (1996) *Toxic Substances: Guidance for the Re-registration of Pesticide Products Containing Glyphosate*, Washington DC: EPA Office of Pesticides, <http://www.epa.gov/> [4]

EPA (1998) *Pesticide Sales 1997*, Washington DC: EPA, <http://www.epa.gov/> [7]

Estok, D., B. Freedman and D. Boyle (1989) 'Effects of the herbicides 2,4-D, glyphosate, hexazinone, and triclopyr on the growth of three species of ectomycorrhizal fungi', *Bulletin of Environmental Contamination and Toxicology*, 42: 835–9. [4]

Ewen, S.W.B. and A. Pusztai (1999) 'Effect of diets containing genetically modified potatoes expressing *Galanthus nivalis* lectin on rat small intestine', *The Lancet*, 354: 1353–4. [8]

FAO (Food and Agriculture Organization) (1997a) *State of the World's Forests*, Rome: FAO. [9]

FAO (1997b) Aquaculture production. *FAO Fisheries Circular*, No. 815, Rome: FAO. [10]

Farrell, A.P., W. Bennett and R.H. Devlin (1996) 'Growth enhanced transgenic salmon can be inferior swimmers', *Canadian Journal of Zoology*, 75: 335–7. [10]

Feldman, J. and T. Stone (1997) 'The development of a comprehensive resistance management plan for potatoes expressing the *Cry3A* endotoxin', in N. Carrozzi and M. Koziel (eds.), *Advances in Insect Control: The Role of Transgenic Plants*, London: Taylor & Francis, pp. 49–61. [4]

Finkel, E. (1999) 'Australian biocontrol beats rabbits, but not rules', *Science*, 285: 1842. [5]

Firbank, L. (ed.) (1999) 'Genetically Modified Crop Farm-Scale Evaluations: First Interim Report, 11 November', London: DETR, <http://www.environment.detr.gov.uk/> [11]

Firbank, L. (ed.) (2000) 'Farm Scale Evaluations of GM Crops: Second Interim Report, 13 June', London: DETR, <http://www.environment.detr.gov.uk/> [11]

Firbank, L.G., A.M. Dewar, M.O. Hill, M.J. May, J.N. Perry, P. Rothery, G.R. Squire and I.P. Woiwod (1999) 'Farm-scale evaluation of GM crops explained', *Nature*, 399: 727–8. [11]

Fisher, R.A. (1930) *The Genetical Theory of Natural Selection*, Oxford: Clarendon Press. [2]

FOE (Friends of the Earth) (1999a) 'GM crops: genetic pollution proved. GM pollen found miles from trial site', Press Release, 29 September, London: FOE. [7]

FOE (1999b) 'Why the GM trials are useless', Press Release, 9 August, London: FOE. [11]

FOE (2000) 'Bee keepers move hives away from GM sites', Press Release, 16 May, London: FOE. [7]

Folmar, L.C., H.O. Sanders and A.M. Julian (1979) 'Toxicity of the herbicide glyphosate and several of its formulations to fish and aquatic invertebrates', *Archives of Environmental Contamination and Toxicology*, 8: 269–78. [4]

Foster, K.R., P. Vecchia and M.H. Repacholi (2000) 'Science and the precautionary principle', *Science*, 288: 979–81. [13]

Fowler, C. and P. Mooney (1990) *The Threatened Gene: Food, Policies and the Loss of Genetic Diversity*, Cambridge: Lutterworth. [12]

Freiburg, B. (1998) 'Is Delta and Pine Land's Terminator Gene a billion dollar discovery?', *Seeds and Crop Digest*, May–June. [8]

Fritz, S.E. and A.J. Lukaszewski (1989) 'Pollen longevity in wheat, rye and triticale', *Plant Breeding*, 102: 31–4. [6]

Gal, S., B. Pisan, T. Hohn, N. Grimsley and B. Hohn (1992) 'Agroinfection of transgenic plants leads to viable Cauliflower Mosaic Virus by intermolecular recombination', *Virology*, 187: 525–33. [4,8]

Genewatch (1999) *Genetically Engineered Oilseed Rape: Agricultural Saviour or New Form of Pollution?*, Buxton: Genewatch. [4]

Gerngross, T.U. and S.C. Slater (2000) 'How green are green plants?', *Scientific American Online*, <http://www.sciam.com/> [4]

Gibbs, G. (1998) 'Devon crop destroyed by gene activists', *Guardian*, 5 August. [6, 11]

Gledhill, M. and P. McGrath (1997) 'Call for a spin doctor', *New Scientist*, 1 November, pp. 4–5. [6]

Gliddon, C.J. (1999) 'Gene flow and risk assessment', in P.J.W. Lutman (ed.), *Gene Flow and Agriculture: Relevance for Transgenic Crops*, Farnham: British Crop Protection Council, pp. 49–56. [6, 11]

Gophen, M., P.B.O. Ochumba and L.S. Kaufman (1995) 'Some aspects of perturbation in the structure and diversity of the ecosystem of Lake Victoria (East Africa)', *Aquatic Living*

Resources, 8: 27–41. [5]

Gould, F. (1998) 'Sustainability of transgenic insecticidal cultivars: integrating pest genetics and ecology', *Annual Review of Entomology*, 43: 701–26. [4]

Gray, A.J. and A.F. Raybould (1998) 'Reducing transgene escape routes', *Nature*, 392: 653–4. [8]

Greenpeace (1999) *Background Briefing on the Biosafety Protocol*, <http://www.greenpeace.com/> [13]

Grossmann, E. and D. Atkinson (eds.) (1985) *The Herbicide Glyphosate*, London: Butterworth. [4]

Gutrich, J.J. and H.W. Whiteman (1998) 'Analysis of the ecological risks associated with genetically engineered macro-organisms', in R.A. Zilinskas and P.J. Balint (eds.), *Genetically Engineered Marine Organisms: Environmental and Economic Risks and Benefits*, pp. 61-93, Norwell, Massachusetts: Kluwer Academic. [2, 10]

Haeckel, E. (1870) 'Über Entwickelungsgang und Aufgabe der Zoologie', *Jenaische Zeitschrift*, 5:353–70. [2]

Hails, R. (2000) 'Genetically modified plants – the debate continues', *Trends in Ecology and Evolution*, 15: 14–18. [2, 6]

Hamilton, W.D. (1963) 'The evolution of altruistic behaviour', *American Naturalist*, 97: 354–6. [2]

Hamilton, W.D. and T.M. Lenton (1998) 'Spora and Gaia: how microbes fly with their clouds', *Ethology, Ecology and Evolution*, 10: 1–16. [3]

Hammock, B. (1991) 'Virus release evaluation', *Nature*, 355: 119. [3]

Hansen Jesse, L.C. and J.J. Obrycki (2000) 'Field deposition of *Bt* transgenic corn pollen: lethal effects on the monarch butterfly', *Oecologia*, 125: 241–8. [7]

Hardin, G. (1960) 'The competitive exclusion principle', *Science*, 131: 1292–7. [2]

Hassan, S.A., F. Bigler, H. Bogenschütz, E. Boller, J. Brun, P. Chiverton, P. Edwards, F. Mansour, E. Naton, P.A. Oomen, W.P.J. Overmeer, L. Polgar, W. Rieckmann, L. Samsøe-Petersen, A. Stäubli, G. Sterk, K. Tavares, J.J. Tuset, G. Viggiani and A.G. Vivas (1988) 'Results of the fourth joint pesticide testing programme carried out by the IOBC/WPRS-Working Group "Pesticides and Beneficial Organisms"', *Journal of Applied Entomology*, 105: 321–9. [4]

Hauser, T.P., R.G. Shaw and H. Østergård (1998a) 'Fitness of F_1 hybrids between weedy *Brassica rapa* and oilseed rape (*B. napus*)', *Heredity*, 81: 429–35. [6]

Hauser, T.P., R.B. Jørgensen and H. Østergård (1998b) 'Fitness of backcross and F_2 hybrids between weedy *Brassica rapa* and oilseed rape (*B. napus*)', *Heredity*, 81: 436–43. [6]

Hew, C.L., G.L. Fletcher and P.L. Davies (1995) 'Transgenic salmon: tailoring the genome for food production', *Journal of Fish Biology*, 47: 1–19. [10]

Hew, C., R. Poon, F. Xiong, S. Gauthier, M. Shears, M. King, P. Davies and G. Fletcher (1999) 'Liver-specific and seasonal expression of transgenic Atlantic salmon harbouring the winter flounder antifreeze protein gene', *Transgenic Research*, 8: 405–14. [10]

Hilbeck, A., M. Baumgartner, P. M. Fried and F. Bigler (1998) 'Effects of transgenic *Bacillus thuringiensis* corn-fed prey on mortality and development time of immature *Chrysoperla carnea* (Neuroptera: Chrysopidae)', *Environmental Entomology*, 27 (2): 480–7. [7]

Hilder, V.A., A.M.R. Gatehouse and D. Boulter (1990) 'Genetic engineering of crops for insect resistance using genes of plant origin', in G.W. Lycett and D. Grierson (eds.), *Genetic Engineering of Crops*, London: Butterworth, pp. 51–66. [4]

Hinchee, M.A.W., S.R. Padgette, G.M. Kishore, X. Delannay and R.T. Fraley (1993) 'Herbicide-Tolerant Crops', in S.D. King and R. Wu (eds.), *Transgenic Plants Vol. 1: Engineering and Utilization*, New York: Academic Press, pp. 243–63. [4]

Hirst, J.M., O.J. Stedman and G.W. Hurst (1967) 'Long distance spore transport: vertical sections of spore clouds over the sea', *Journal of General Microbiology*, 48: 357–77. [9]

Hjelmroos, M. (1991) 'Evidence of long-distance transport of *Betula* pollen', *Grana*, 30: 215–28. [9]

Ho, M.-W. (1998) *Genetic Engineering: Dreams or Nightmare. The Brave New World of Bad Science and Big Business*, Bath: Gateway Books. [11]

Hoffmann, T., C. Golz and O. Schieder (1994) 'Foreign DNA sequences are received by a wild-type strain of *Aspergillus niger* after co-culture with transgenic higher plants', *Current Genetics*, 27: 70–6. [6]

Holm, L.G., D.L. Plucknett, J.V. Pancho and J.P. Harberger (1977) *The World's Worst Weeds: Distribution and Biology*, Honolulu: University Press of Hawaii. [12]

Holmes, B. (1995) 'Oaks favour far-flung fathers', *New Scientist*, 26 August, p. 14. [9]

Holmes, B. (1998) 'Gene crop axed', *New Scientist*, 10 January, p. 12. [4]

Holmes, M.T., E.R. Ingham, J.D. Doyle and C.W. Hendricks (1999) 'Effect of *Klebsiella planticola* SDF20 on soil foodweb organisms and wheat growth in sandy soil', *Applied Soil Ecology*, 11: 67–78. [3]

Holtby, L B. and S.J. Baillie (1987) 'Effects of the herbicide Roundup on coho salmon fingerlings in an over-sprayed tributary of Carnation Creek, British Columbia', in P.E. Reynolds (ed.), *Proceedings of the Carnation Creek Herbicide Workshop*, December 7–10, 1987, British Columbia: Forest Pest Management Institute, pp. 273–85. [4]

Horvath, L. and L. Orban (1995) 'Genome and gene manipulation in the common carp', *Aquaculture*, 129: 157–81. [10]

Hoy, M.A. (2000) 'Deploying transgenic arthropods in pest management programs: risks and realities', in A.M. Handler and A.J. James (eds.), *Insect Transgenesis: Methods and Applications*, Boca Raton, Florida: CRC Press, pp. 335–67. [11]

Hoyle, B. (1999) 'Canadian farmers seek compensation for "genetic pollution"', *Nature Biotechnology*, 17: 747–8. [6]

Huang, E., L.L. Buschman, R.A. Higgins and W.H. McGaughey (1999) 'Inheritance of resistance to *Bacillus thuringiensis* toxin (Dipel ES) in the European corn borer', *Science*, 284: 965–7. [4]

Hunt, L. (1998) 'Send in the clouds', *New Scientist*, 30 May, pp. 29–33. [3]

Hutchinson, G.E. (1957) 'Concluding remarks', *Cold Spring Harbour Symposium on Quantitative Biology*, 22: 415–27. [2]

Ingham, E. (1998) 'Good intentions and engineering organisms that kill wheat', *Synthesis/Regeneration*, 18, <http://www.greens.org/s-r/18/18-14.html> [3]

Iyengar, A., F. Müller and N. MacLean (1996) 'Regulation and expression of transgenes in fish – a review', *Transgenic Research*, 5: 147–66. [10]

Jakowitsch, J., M. Mette, J. van der Winden, M. Matzke and A. Matzke (1999) 'Integrated pararetroviral sequences define a unique class of dispersed repetitive DNA in plants', *Proceedings of the National Academy of Sciences, USA*, 96: 13241–6. [8]

James, C. (2000) *Global Status of Commercialized Transgenic Crops: 2000*, ISAAA Briefs No. 21: Preview, Ithaca, NY: International Service for the Acquisition of Agri-biotech Applications. [1, 11, 12]

Jayaraman, K.S. (2001) 'Illicit GM cotton sparks corporate fury', *Nature*, 413: 555. [6]

Jones, M.D. and L.C. Newell (1948) 'Longevity of pollen and stigmas of grasses: buffalograss, *Buchloe dactyloedees* (NUTT) Engelm, and corn, *Zea mays* L.', *Journal of the American Society of Agronomy*, 40 (3): 195–204. [6]

Jørgensen, R.B. (1999) 'Gene flow from oilseed rape (*Brassica napus*) to related species', in P.J.W. Lutman (ed.), *Gene Flow and Agriculture: Relevance for Transgenic Crops*, Farnham:

British Crop Protection Council, pp. 117–23. [6]

Jørgensen, R. B. and B. Andersen (1994) 'Spontaneous hybridization between oilseed rape, *Brassica napus*, and weedy *B. Campestris* (Brassicaceae): A risk of growing genetically modified oilseed rape', *American Journal of Botany*, 81: 1620–6. [6]

Jørgensen, R.B., B. Andersen, L. Landbo and T.R. Mikkelsen (1996) 'Spontaneous hybridization between oilseed rape (*Brassica napus*) and weedy relatives', *Acta Horticulturae*, 407: 193–200. [6]

Jørgensen, R.B., B. Andersen, T.P. Hauser, L. Landbo, T.R. Mikkelsen and H. Østergård (1998) 'Introgression of crop genes from oilseed rape (*Brassica napus*) to related weed species – an avenue for the escape of engineered genes', *Acta Horticulturae*, 459: 211–17. [6]

Juma, C. (1989) *The Gene Hunters: Biotechnology and the Scramble for Seeds*, London: Zed Books. [8, 12]

Kapuscinski, A.R. and E.M. Hallerman (1990) 'Transgenic fish and public policy: anticipating environmental impacts of transgenic fish', *Fisheries*, 15 (1): 2–11. [10]

Keeler, K.H., C.E. Turner and M.R. Bolick (1996) 'Movement of crop transgenes into wild plants', in S.O. Duke (ed.), *Herbicide-Resistant Crops: Agricultural, Economic, Environmental, Regulatory, and Technological Aspects*, Boca Raton, Florida: CRC Press, pp. 303–30. [6]

Kelso, P. (2000) 'Greenpeace wins key GM case', *Guardian*, 21 September. [11]

Kiernan, V. (1993) 'US counts cost of alien invaders', *New Scientist*, 23 October, p. 9. [5]

Kiernan, V. (1996) 'Yes, we have no vaccinating bananas', *New Scientist*, 21 September, p. 6. [4, 9]

King, D. (2000) 'Butterfly study causes furore', *GenEthics News*, 30/31: 9. [7]

King, J. (1996) 'Could transgenic supercrops one day breed superweeds?', *Science*, 274: 180–1. [6]

Kleiner, K. (1997a) 'Fields of genes', *New Scientist*, 16 August, p. 4. [4]

Kleiner, K. (1997b) 'Human harvest festival', *New Scientist*, 12 July, p. 17. [4]

Kleiner, K. (2000) 'Unfit for humans corn has been contaminated with a potentially harmful protein', *New Scientist*, 2 December, p. 11. [6]

Kleiner, K. (2001) 'Victory for Monsanto: If modified plants contaminate crops it could cost you dear', *New Scientist*, 7 April, p. 13. [6]

Klinger, T. and N.C. Ellstrand (1994) 'Engineered genes in wild populations: fitness of weed–crop hybrids of *Raphanus sativus*', *Ecological Applications*, 4: 117–20. [6]

Klinger, T., P.E. Arriola and N.C. Ellstrand (1992) 'Crop–weed hybridization in radish (*Raphanus sativus* L.): Effects of distance and population size', *American Journal of Botany*, 79: 1431–5. [6]

Knibb, W. (1997) 'Risk from genetically engineered and modified marine fish', *Transgenic Research*, 6: 54–67. [10]

Koechlin, F. (2000) *The 'Golden Rice': A Big Illusion?*, <http://www.blauen-institute.ch/> [12]

Koga, Y., T. Akihama, H. Fujimaki and M. Yokoo (1971) 'Studies on the longevity of pollen grains of rice *Oryza sativa* L. I. Morphology changes of pollen grains after shedding', *Cytologia*, 36: 104–10. [6]

Kota, M., H. Daniell, S. Varma, F. Garczynski, F. Gould and W.J. Moar (1999) 'Overexpression of the *Bacillus thuringiensis* Cry2A protein in chloroplasts confers resistance to plants against susceptible and *Bt*-resistant insects', *Proceedings of the National Academy of Sciences, USA*, 96: 1840–5. [8]

Krebs, C.J. (1985) *Ecology: the Experimental Analysis of Distribution and Abundance*, 3rd Edition, London: Harper and Row. [2]

Kryder, R.D., S.P. Kowalski and A.F. Krattiger (2000) *The Intellectual and Technical Property*

Components of pro-Vitamin A Rice (GoldenRice™): A Preliminary Freedom-to-operate Review, ISAAA Briefs No. 20, Ithaca, New York: ISAAA. [4]

Kwain, W. and A.H. Lawrie (1981) 'Pink salmon in the Great Lakes', *Fisheries*, 6 (2): 2–6. [10]

Langelle, O. (2001) 'From native forest to frankenforest', in B. Tokar (ed.), *Redesigning Life?: The Worldwide Challenge to Genetic Engineering*, London: Zed Books, pp. 111–25. [9]

Lanner, R.M. (1966) 'Needed: a new approach to the study of pollen dispersal', *Silvae Genetica*, 15: 50–2. [6]

Lappé, M. and B. Bailey (1999) *Against the Grain: The Genetic Transformation of Global Agriculture*, London: Earthscan. [4]

Leahy, S. (1998) 'First GM tolerance transfer is feared in Canadian OSR', *Farmers Weekly*, 15 January. [6]

Lean, G. (2000) 'Commons misled over GM bungle', *Independent on Sunday*, 21 May. [6]

Lean, G., V. Angres and L. Jury (2000) 'GM genes "can spread to people and animals"', *Independent*, 28 May. [7]

Leapman, M. (1999) 'Tommy Archer got off – but will Peter Melchett?', *Independent on Sunday*, 7 November. [11]

LePage, M. (2000) 'Seeds of dispute', *New Scientist*, 23 December, pp. 22–3. [6]

Leslie, M. (2001) 'Tales of the sea', *New Scientist*, 27 January, pp. 32–5. [13]

Levin, D.A. (1984) 'Immigration in plants: an exercise in the subjunctive', in R. Dirzo and J. Sarukhan (eds.), *Perspectives on Plant Population Ecology*, Sunderland: Sinauer, pp. 242–60. [6]

Levin, D.A. and H.W. Kerster (1974) 'Gene flow in seed plants', *Evolutionary Biology*, 7: 139–220. [6]

Levy, S.B. and B.M. Marshall (1988) 'Genetic transfer in the natural environment', in M. Sussman, C.H. Collins, F.A. Skinner and D.E. Stewart-Tull (eds.), *The Release of Genetically-Engineered Micro-organisms*, London: Academic Press, pp. 61–76. [3]

Lewis, M.A., G. Schmitz, P. Kareiva and J.T. Trevors (1996) 'Models to examine containment and spread of genetically engineered microbes', *Molecular Ecology*, 5: 165–75. [3]

Lewis, R. and B.A. Palevitz (1999) 'GM crops face heat of debate', *The Scientist*, 13 (20): 1, 8–9. [7]

Lewontin, R.C. (1993) *The Doctrine of DNA: Biology as Ideology*, Harmondsworth: Penguin. [8]

Linder, C.R., I. Taha, G.J. Seiler, A.A. Snow and L.H. Rieseberg (1998) 'Long-term introgression of crop genes into sunflower populations', *Theoretical and Applied Genetics*, 96: 339–47. [6]

Lindow, S.E. and N.J. Panopoulos (1988) 'Field tests of recombinant ice minus *Pseudomonas syringae* for biological frost control in potato', in M. Sussman, C.H. Collins, F.A. Skinner and D.E. Stewart-Tull (eds.), *Proceedings of the First International Conference on the Release of Genetically-Engineered Micro-organisms*, New York: Academic Press, pp. 121–38. [3]

Liong, P.C., W.P. Hamzah, V. Murugan (1988) 'Toxicity of some pesticides towards freshwater fishes', *Malaysian Agricultural Journal*, 54 (3): 147–56. [4]

Liu, Y.-B., B.E. Tabashnik, T.J. Dennehy, A.L. Patin and A.C. Bartlett (1999) 'Development time and resistance to Bt crops', *Nature*, 400: 519. [4]

Llewellyn, D. and G. Fitt (1996) 'Pollen dispersal from two fields of transgenic cotton in the Namoi valley, Australia', *Molecular Breeding*, 2: 157–66. [6]

Lockhart, B., J. Menke, G. Dahal and N. Olszwewski (2000) 'Characterization and genomic analysis of tobacco vein clearing virus, a plant pararetrovirus that is transmitted vertically and related to sequences integrated in the host genome', *Journal of General Virology*, 81: 1579–85. [8]

Lorenz, M.G. and W. Wackernagel (1993) 'Bacterial gene transfer in the environment', in K. Wöhrmann and J. Tomiuk (eds.), *Transgenic Organisms: Risk Assessment of Deliberate Release*, Basel: Birkhäuser Verlag, pp. 43–64. [3]

Lorenz, M.G. and W. Wackernagel (1994) 'Bacterial gene transfer by natural genetic transformation', *Microbiological Reviews*, 58 (3): 563–602. [3]

Losey, J.E., L.S. Rayor and M.E. Carter (1999) 'Transgenic pollen harms monarch larvae', *Nature*, 399: 214. [7]

Lovelock, J. (1988) *The Ages of Gaia: A Biography of Our Living Earth*, Oxford: Oxford University Press. [2]

McArthur, R.H. and E.O. Wilson (1963) 'An equilibrium theory of insular zoogeography', *Evolution*, 17 (4): 373–87. [2]

McBride, K.E., Z. Svab, D.J. Schaaf, P.S. Hogan, D.M. Stalker and P. Maliga (1995) 'Amplification of a chimeric *Bacillus* gene in chloroplasts leads to an extraordinary level of an insecticidal protein in tobacco', *BioTechnology*, 13: 362–5. [8]

McCarthy, M. (2000) 'Honey has been contaminated by GM crops, claims Friends of the Earth', *Independent*, 17 May. [7]

McGreal, C. (2001) 'Shamed and humiliated – the drug firms back down', *Guardian*, 19 April. [12]

Mack, R.N. (1991) 'The commercial seed trade: an early disperser of seeds in the United States', *Economic Botany*, 45: 257–73. [5]

Mack, R.N. (1996) 'Predicting the identity and fate of plant invaders: emergent and emerging approaches', *Biological Conservation*, 78: 107–21. [5]

MacKenzie, D. (1999) 'Gut reaction', *New Scientist*, 30 January, p. 4. [8]

MacKenzie, D. (2000) 'Stray genes highlight superweed danger', *New Scientist*, 21 October, p. 6. [6]

MacKenzie, D. (2001) 'This means war: alien invaders are decimating native wildlife all around the world', *New Scientist*, 24 March, p. 12. [5]

Maclean, N. (1994) 'Transgenic animals in perspective', in N. Maclean (ed.), *Animals with Novel Genes*, Cambridge: Cambridge University Press, pp. 1–20. [1]

Maclean, N. (1998) 'Genetic manipulation of farmed fish', in K.D. Black and A.D. Pickering (eds.), *Biology of Farmed Fish*, Sheffield: Sheffield Academic Press, pp. 327–54. [10]

Maclean, N. and A. Rahman (1994) 'Transgenic fish', in N. Maclean (ed.), *Animals With Novel Genes*, Cambridge: Cambridge University Press, pp. 63–105. [10]

MAFF (1999) 'Scheme to prevent injurious cross pollination of certain crops grown in North Essex', News Release, 7 May, Cambridge: MAFF. [6]

Malakoff, D. (1999) 'Fighting fire with fire', *Science*, 285: 1841–3. [5]

Malcolm, S.B., B.J. Cockrell and L.P. Brower (1993) 'Spring recolonization of North America by the monarch butterfly: successive brood or single sweep migration', in S.B. Malcolm and M.P. Zalucki (eds.), *Biology and Conservation of the Monarch Butterfly*, Los Angeles: Natural History Museum of Los Angeles County, pp. 253–67. [7]

Mandrioli, P., M.G. Negrini, G. Cesari and G. Morgan (1984) 'Evidence for long range transport of biological and anthropogenic aerosol particles in the atmosphere', *Grana*, 23: 43–53. [9]

Marchant, J. (2000) 'Life in the clouds', *New Scientist*, 26 August, p. 4. [3]

Margulis, L. (1993) *Symbiosis in Cell Evolution*, 2nd Edition, San Francisco: Freeman. [8]

Martinez, R., M.P. Estrada, J. Berlanga, I. Guillém, O. Hernández *et al.* (1996) 'Growth enhancement in transgenic tilapia by ectopic expression of tilapia growth hormone', *Molecular Marine Biology and Biotechnology*, 5: 62–70. [10]

Martinez, T.T. and K. Brown (1991) 'Oral and pulmonary toxicology of the surfactant used in Roundup herbicide', *Proceedings of the Western Pharmacological Society*, 34: 43–6. [4]

Masood, E. (1998) 'Organic farmer takes gene battle to court', *Nature*, 394: 8. [6]

Matzke, M.A., M. Primig, J. Trnovsky and A.J.M. Matzke (1989) 'Reversible methylation and inactivation of marker genes in separately transformed tobacco plants', *EMBO Journal*, 8: 643–9. [11]

Meek, J. (2001) 'Scientists create killer moth to control pests', *Guardian*, 5 March. [11]

Meikle, J. (1998) 'Farmer challenges gene crops', *Guardian*, 10 July. [6]

Meikle, J. (2000a) 'Imported seeds tainted by GM', *Guardian*, 18 May. [6]

Meikle, J. (2000b) 'Company to compensate farmers for GM muddle', *Guardian*, 6 June. [6]

Meikle, J. (2000c) 'GM "pollution" unstoppable, says Meacher', *Guardian*, 14 June. [6]

Meikle, J. (2000d) 'Dairy forces farmers to abandon GM crop trial', *Guardian*, 4 May. [11]

Meikle, J. (2000e) 'GM genes can jump species, says expert', *Guardian*, 29 May. [7]

Meinsz, A. (2001) *Killer Algae*, Chicago: University of Chicago Press. [5]

Mendelson, J. (1998) 'Roundup: the world's biggest selling herbicide', *The Ecologist*, 28: 270–75. [4]

Metz, P.L.J., E. Jacobsen and W.J. Stiekema (1997) 'Occasional loss of expression of phosphinothricin tolerance in sexual offspring of transgenic oilseed rape (*Brassica napus* L.)', *Euphytica*, 98: 189–96. [11]

Mikkelsen, T.R., B. Andersen and R.B. Jørgensen (1996) 'The risk of crop transgene spread', *Nature*, 380: 31. [6]

Millar, S. (2001) 'How the king of fish is being farmed to death', *Observer*, 7 January. [10]

Mitchell, D.G., P.M. Chapman and T.J. Long (1987) 'Acute toxicity of Roundup and Rodeo herbicides to rainbow trout, chinook and coho salmon', *Bulletin of Environmental Contamination and Toxicology*, 39: 1028–35. [4]

Monbiot, G. (1999) 'Disruptive protest is a civic duty', *Guardian*, 19 August. [11]

Monsanto (1999) *Butterflies and Bt Corn Pollen: Lab Research and Field Realities*, St. Louis, Missouri: Monsanto. [7]

Monsanto (2000a) *Glyphosate*, St. Louis, Missouri: Monsanto, <http://www.monsanto.com/> [4]

Monsanto (2000b) *Monsanto Offers Patent Rights to Golden Rice*, St. Louis, Missouri: Monsanto, <http://www.monsanto.com/> [4]

Moscardi, F. (1999) 'Assessment of the application of Baculovirus for control of Lepidoptera', *Annual Review of Entomology*, 44: 257–89. [3]

Müller, F., Z. Ivics, F. Erdelyi, T. Papp, L. Varadi, L. Horvath, N. Maclean and L. Orban (1992) 'Introducing foreign genes into fish eggs with electroporated sperm as a carrier', *Molecular Marine Biology and Biotechnology*, 1: 276–81. [10]

Nap, J.P., J. Bijvoet and W.J. Stiekema (1992) 'Biosafety of kanamycin-resistant transgenic plants', *Transgenic Research*, 1: 239–49. [8]

Ndowora, T., G. Dahal, D. LaFleur, G. Harper, R. Hull, N. Olszewski and B. Lockhart (1999) 'Evidence that badnavirus in *Musa* can originate from integrated pararetroviral sequences', *Virology*, 255: 214–20. [8]

Nicholl, S.T. (1994) *An Introduction to Genetic Engineering*, Cambridge: Cambridge University Press. [1]

Niiler, E. (1999) 'GM corn poses little threat to monarch', *Nature Biotechnology*, 17: 1154. [7]

Nordlee, J.A., S.L. Taylor, J.A. Townsend, L.A. Thomas and R.K. Bush (1996) 'Identification of a brazil-nut allergen in transgenic soybeans', *New England Journal of Medicine*, 334: 688–92. [6]

Nordlund, D.A. and W.J. Lewis (1976) 'Terminology of chemical releasing stimuli in intraspecific and interspecific interactions', *Journal of Chemical Ecology*, 2: 211–20. [7]

Nottingham, S.F. (1998) *Eat Your Genes: How Genetically Modified Food is Entering Our Diet*, London: Zed Books, pp. 129–43. [4]

O'Morchoe, S.B., O. Ogunseitan, G.S. Sayler and R.W. Miller (1988) 'Conjugal transfer of R68.45 and FP5 between *Pseudomonas aeruginosa* strains in a freshwater environment', *Applied and Environmental Microbiology*, 54: 1923–9. [3]

Odum, E.P. (1985) 'Biotechnology and the biosphere', *Science*, 27 September, p. 1338. [3]

Oka, H.I. (1988) *Origins of Cultivated Rice*, Tokyo: Japanese Science Society Press. [6]

Oka, H.I. and W.T. Chang (1961) 'Hybrid swarms between wild and cultivated rice species, *Oryza perennis* and *O. sativa*', *Evolution*, 15: 326–39. [6]

Oren, R., D.S. Ellsworth, K.H. Johnsen, N. Phillips, B.E. Ewers, C. Maler, K.V.R. Schäfer, H. McCarthy, G. Hendrey, S.G. McNulty and G.G. Katul (2001) 'Soil fertility limits carbon sequestration by forest ecosystems in a CO_2-enriched atmosphere', *Nature*, 411: 469–72. [9]

Ow, D.W. (2001) 'The right chemistry for marker gene removal?', *Nature Biotechnology*, 19: 115. [8]

Owen, D.F. (1974) *What is Ecology?*, Oxford: Oxford University Press. [2]

Owusu, R.A. (1999) *GM Technology in the Forest Sector: A Scoping Study for WWF*, Washington DC: World Wildlife Foundation. [9]

Oxtoby, E. and M.A. Hughes (1990) 'Engineering herbicide tolerance into crops', *Trends in Biotechnology*, 8: 61–5. [4, 5]

Pearce, D. and R. Tinch (1998) 'The true price of pesticides', in W. Vorley and D. Keeney (eds.), *Bugs in the System: Redesigning the Pesticide Industry for Sustainable Agriculture*, London: Earthscan, pp. 50–93. [12]

Pearce, F. and S. Padulosi (2000) 'Ripe for revival', *New Scientist*, 2 September, pp. 42–5. [12]

Pearce, P. (2001) 'An ordinary miracle', *New Scientist*, 3 February, pp. 16–17. [12]

Peferoen, M. (1997) 'Progress and prospects for field use of *Bt* genes in crops', *Trends in Biotechnology*, 15: 173–7. [4]

Peña, L., M. Martín-Trillo, J. Juárez, J.A. Pina, L. Navarro and J.M. Martínez-Zapater (2001) 'Constitutive expression of *Arabidopsis* LEAFY or APETALAI genes in Citrus reduces their generation time', *Nature Biotechnology*, 19: 263–7. [9]

Pham-Delegue, M.H., A.L. Picard-Nizou, G. Arnold, R. Grison, A. Toppan, L. Olsen and C. Masson (1992) 'Impact of genetically modified rapeseed on insect pollinators (honeybees)', in R. Casper and J. Landsmann (eds.), *Proceedings of the Second International Symposium on the Biosafety of Field Tests of Genetically Modified Plants and Microorganisms*, Goslar: BBA Braunschweig. [7]

Picard-Nizou, A.L., R. Grison, L. Olsen, C. Pioche, G. Arnold and M.H. Pham-Delegue (1997) 'Impact of protein used in plant genetic engineering: Toxicity and behavioral study in the honeybee', *Journal of Economic Entomology*, 90: 1710–16. [7]

Pielou, E.C. (1974) *Population and Community Ecology*, New York: Gordon & Breach. [2]

Pilcher, C.D., J.J. Obrycki, M.E. Rice and C.S. Lewis (1997) 'Preimaginal development, survival, and field abundance of insect predators on transgenic *Bacillus thuringiensis* corn', *Environmental Entomology*, 26: 446–54. [7]

Pimentel, D. (2001) *Ecological Integrity*, New York: Island Press. [10]

Pimentel, D.S. and P.H. Raven (2000) '*Bt* corn pollen impacts on nontarget Lepidoptera: Assessment of effects in nature', *Proceedings of the National Academy of Sciences, USA*, 97 (15): 8198–9. [7]

Pimm, S.L. (1982) *Food Webs*, London: Chapman & Hall. [2]

Poirier, Y. (1999) 'Green chemistry yields a better plastic', *Nature Biotechnology*, 17: 1011–16. [4]

Poirier, Y., D.E. Dennis, K. Klomparins and C. Somerville (1992) 'Polyhydroxybutyrate, a biodegradable thermoplastic, produced in transgenic plants', *Science*, 256: 520–622. [4]

Potrykus, I. (2001) 'Response to "Genetically engineered rice is 'fool's gold' "', <http://www.biotech-info.net/> [12]

Pretty, J. (1995) *Regenerating Agriculture: Policies and Practice for Sustainability and Self-Reliance*, London: Earthscan. [12]

Pretty, J. (1998) *The Living Land: Agriculture, Food and Community Regeneration in Rural Europe*, London: Earthscan. [12]

Pretty, J. (2001) 'Against the grain', *Guardian (Society)*, 17 January. [12]

Prince, R.C. (1997) 'Bioremediation of marine oil spills', *Trends in Biotechnology*, 15(5): 158–60. [3]

Purseglove, J.W. (1972) *Tropical Crops: Monocotyledons*, London: Longman Group. [6]

Pursel, V.G. (1997) 'Techniques and problems in producing transgenic pigs', in L.M. Houbedine (ed.), *Transgenic Animals – Generation and Use*, Amsterdam: Harwood Academic, pp. 15–17. [10]

Quinn, S. (1999) 'Six arrests after GM crop attack', *Guardian*, 19 July. [11]

Radford, T. (1999) 'Pollen from GM maize shown to kill butterflies', *Guardian*, 20 May. [7]

Radford, T. (2000) 'GM crops threaten skylarks', *Guardian*, 1 September. [7]

RAFI (Rural Advancement Foundation International) (1997) 'Bioserfdom: technology, intellectual property and the erosion of farmers' rights in the industrialized world', *RAFI Communiqué*, March/April, <http://www.rafi.org/> [8]

RAFI (1998) 'Terminator technology targets farmers', *RAFI Communiqué*, March/April, <http://www.rafi.org/> [8, 12]

RAFI (1999) 'The gene giants: update on consolidation in the life industry', *RAFI Communiqué*, March, <http://www.rafi.org/> [12]

RAFI (2001a) 'Monsanto's "submarine patent" torpedoes ag biotech', *RAFI News Release*, April, <http://www.rafi.org/> [8]

RAFI (2001b) 'New terminator patent goes to Syngenta', *RAFI News Release*, March, <http://www.rafi.org/> [8]

RAFI (2001c) 'The ETC Century: Erosion, technological transformation, and corporate concentration in the 21st century', *RAFI Communiqué*, February, <http://www.rafi.org/> [12]

Rahman, M.A., R. Mah, H. Ayad, A. Smith and N. Maclean (1998) 'Expression of a novel piscine growth hormone gene results in growth enhancement in transgenic tilapia (*Oreochromis niloticus*)', *Transgenic Research*, 7: 357–69. [10]

Ramsay, G., C.E. Thompson, S. Neilson and G. Mackay (1999) 'Honeybees as vectors of GM oilseed rape pollen', in P.J.W. Lutman (ed.), *Gene Flow and Agriculture: Relevance for Transgenic Crops*, Farnham: British Crop Protection Council, pp. 209–14. [6, 7]

Raybould, A.F. and R.T. Clarke (1999) 'Defining and measuring gene flow', in P.J.W. Lutman (ed.), *Gene Flow and Agriculture: Relevance for Transgenic Crops*, Farnham: British Crop Protection Council, pp. 41–8. [6]

Raybould, A.F. and A.J. Gray (1993) 'Genetically modified crops and hybridization with wild relatives: A UK perspective', *Journal of Applied Ecology*, 30: 199–219. [6]

Raybould, A.F. and A.J. Gray (1994) 'Will hybrids of genetically modified crops invade natural communities?', *Trends in Ecology and Evolution*, 9: 85–9. [6]

Rieseberg, C.H., M.J. Kim and G.J. Seiler (1999) 'Introgression between the cultivated sunflower and a sympatric wild relative, *Helianthus petiolaris* (Asteraceae)', *International Journal of Plant Science*, 160: 102–8. [6]

Ricklefs, R.E. (1973) *Ecology*, New York: Nelson. [2]

Richards, A.J. (1986) *Plant Breeding Systems*, London: George Allen & Unwin. [6]

Rifkin, J. (1998) *The Biotech Century*, London: Victor Gollancz, pp. 67–115. [12]

Rissler, J. and M. Mellon (1996) *The Ecological Risks of Engineered Crops*, Cambridge, Massachusetts: MIT Press, pp. 71–109. [11]

Robinson, D.J. (1996) 'Environmental risk assessment of releases of transgenic plants containing virus-derived inserts', *Transgenic Research*, 5: 359–62. [4]

Sample, I. (2000) 'Modified crops could corrupt weedy cousins', *New Scientist*, 15 July, p. 6. [6]

Sattler, B., H. Puxbaum and R. Psenner (2001) 'Bacterial growth in supercooled cloud droplets', *Geophysical Research Letters*, 28: 239–42. [3]

Sawada, Y.Y., Y. Nagai, M. Ueyama and I. Yamamoto (1988) 'Probable toxicity of surface-active agents in commercial herbicide containing glyphosate', *The Lancet*, 8580: 229. [4]

Saxena, D., S. Flores and G. Stotzky (1999) 'Insecticidal toxin in root exudates from *Bt* corn', *Nature*, 402: 480. [7]

Scheffler, J.A. and P.J. Dale (1994) 'Opportunities for gene transfer from transgenic oilseed rape (*Brassica napus*) to related species', *Transgenic Research*, 3: 263–78. [6]

Schlesinger, W.H. and J. Lichter (2001) 'Limited carbon storage in soil and litter of experimental forest plots under increased atmospheric CO_2', *Nature*, 411: 466–9. [9]

Schubbert, R., C. Lettmann and W. Doerfler (1994) 'Ingested foreign (phage M13) DNA survives transiently in the gastrointestinal tract and enters the bloodstream of mice', *Molecular and General Genetics*, 242: 495–504. [8]

Schubbert, R., D. Renz, B. Schmitz and W. Doerfler (1997) 'Foreign (M13) DNA ingested by mice reaches peripheral leukocytes, spleen and liver via the intestinal wall mucosa and can be covalently linked to mouse DNA', *Proceedings of the National Academy of Sciences, USA*, 94: 961–6. [8]

Schubbert, R., U. Hohlweg, D. Renz and W. Doerfler (1998) 'On the fate of orally ingested foreign DNA in mice: chromosomal association and placental transmission to the fetus', *Molecular and General Genetics*, 259: 569–76. [8]

Schuler, T.J., G.M. Poppy, B.R. Kerry and I. Denholm (1998) 'Insect-resistant transgenic plants', *Trends in Biotechnology*, 16: 168–75. [4]

Schuler, T.J., R.P.J. Potting, I. Denholm and G.M. Poppy (1999) 'Parasitoid behaviour and *Bt* plants', *Nature*, 400: 825. [7]

Scott, S.E. and M.J. Wilkinson (1999) 'Low probability of chloroplast movement from oilseed rape (*Brassica napus*) into wild *Brassica rapa*', *Nature Biotechnology*, 17: 390–2. [8]

Scupham, A.J., A.H. Bosworth, W.R. Ellis, T.J. Wacek, K.A. Albrecht and E.W. Triplett (1996) 'Inoculation with *Sinorhizobium meliloti* RMBPC-2 increases alfalfa yield compared with inoculation with a nonengineered wild-type strain', *Applied and Environmental Microbiology*, 62: 4260–2. [3]

Seefeldt, S.S., F.L. Young, R. Zenetra and S.S. Jones (1999) 'The production of herbicide-resistant jointed goatgrass (*Aegilops cylindrica*) X wheat (*Triticum aestivum*) hybrids in the field by natural hybridization', in P.J.W. Lutman (ed.), *Gene Flow and Agriculture: Relevance for Transgenic Crops*, Farnham: British Crop Protection Council, pp. 159–63. [6]

Sen, A. (1981) *Poverty and Famines: An Essay on Entitlement and Deprivation*, Oxford: Clarendon Press. [12]

Senior, I.J. and P.J. Dale (1999) 'Molecular aspects of multiple transgenes and gene flow in crops and wild relatives', in P.J.W. Lutman (ed.), *Gene Flow and Agriculture: Relevance for Transgenic Crops*, Farnham: British Crop Protection Council, pp. 225–31. [11]

Servizi, J.A., R.W. Gordon and D.W. Martens (1987) 'Acute toxicity of Garlon 4 and Roundup herbicides to salmon, *Daphnia*, and trout', *Bulletin of Environmental Contamination and Toxicology*, 39: 15–22. [4]

Shand, H. (2001) 'Gene Giants: Understanding the "Life Industry"', in B. Tokar (ed.),

Redesigning Life?: The Worldwide Challenge to Genetic Engineering, London: Zed Books, pp. 222–37. [12]

Sharp, R., A.G. O'Donull, H.G. Gilbert and G.P. Hazelwood (1992) 'Growth and survival of genetically manipulated *Lactobacillus plantarum* in silage', *Applied and Environmental Microbiology*, 58: 1517–22. [3]

Shears, M.A., G.L. Fletcher, C.L. Hew, S. Gauthier and P.L. Davies (1991) 'Transfer, expression and stable inheritance of antifreeze protein genes in Atlantic salmon (*Salmo salar*)', *Marine Molecular Biology and Biotechnology*, 1: 58–63. [10]

Shelton, A.M. and R. Roush (1999) 'False reports and the ears of men', *Nature Biotechnology*, 17: 832. [7]

Shiva, V. (1993) *Monocultures of the Mind: Perspectives on Biodiversity and Biotechnology*, London: Zed Books. [12]

Shiva, V. (1998) *Biopiracy: The Plunder of Nature and Knowledge*, Totnes: Green Books. [12]

Shiva, V. (2001a) 'Biopiracy: The theft of knowledge and resources', in B. Tokar (ed.), *Redesigning Life?: The Worldwide Challenge to Genetic Engineering*, pp. 283–9, London: Zed Books. [12]

Shiva, V. (2001b) 'Genetically engineered 'Vitamin A rice': A blind approach to blindness prevention', in B. Tokar (ed.), *Redesigning Life?: The Worldwide Challenge to Genetic Engineering*, London: Zed Books, pp. 40–3. [12]

Short, K.A., R.J. Seidler and R.H. Olsen (1990) 'Survival and degradative capacity of *Pseudomonas putida* induced or constitutively expressing plasmid-mediated degradation of 2,4-dichlorophenoxyacetate in soil', *Canadian Journal of Microbiology*, 36: 821–6. [3]

Simkiss, K. (1994) 'Transgenic birds', in N. Maclean (ed.), *Animals with Novel Genes*, Cambridge: Cambridge University Press, pp. 106–37. [11]

Skogsmyr, I. (1994) 'Gene dispersal from transgenic potatoes to conspecifics: A field trial', *Theoretical and Applied Genetics*, 88: 770–4. [6]

Snow, A.A. and R.B. Jørgensen (1999) 'Fitness costs associated with transgenic glufosinate tolerance introgression from *Brassica napus* spp. *oleifera* (oilseed rape) into weedy *Brassica rapa*', in P.J.W. Lutman (ed.), *Gene Flow and Agriculture: Relevance for Transgenic Crops*, Farnham: British Crop Protection Council, pp. 137–42. [6]

Snow, A.A. and P. Morán Palma (1997) 'Commercialization of transgenic plants: potential ecological risks', *BioScience*, 47: 86–96. [6, 11]

Southwood, T.R.E. (1978) *Ecological Methods*, 2nd Edition, London: Chapman & Hall. [2]

Souzha-Machado, V., J.D. Bandeen, G.R. Stephenson and P. Lavigne (1978) 'Uniparental inheritance of chloroplast atrazine tolerance in *Brassica campestris*', *Canadian Journal of Plant Science*, 58: 977–81. [8]

Springett, J.A. and R.A.J. Gray (1992) 'Effect of repeated low doses of biocides on the earthworm *Aporrectodea caliginosa* in laboratory culture', *Soil Biology and Biochemistry*, 24: 1739–44. [4]

Squire, G., N. Augustin, J. Bown, J.W. Crawford, G. Dunlop, J. Graham, J.R. Hillman, B. Marshall, D. Marshall, G. Ramsay, D.J. Robinson, J. Russell, C. Thompson and G. Wright (1999) 'Gene flow in the environment – genetic pollution?', Scottish Crop Research Institute Annual Report for 1998/1999, <http://www.scri.sari.ac.uk/Documents/AnnReps/AnnRep.htm> [6]

Steinbrecher, R.A. and P.R. Mooney (1998) 'Terminator technology: the threat to world security', *The Ecologist*, 28 (5): 276–9. [8]

Stewart, C.N. Jr., J.N. All, P.L. Raymer and S. Ramachandran (1997) 'Increased fitness of transgenic insecticidal rapeseed under insect selection pressure', *Molecular Ecology*, 6: 773–9. [5]

Stewart, C.N. Jr. and S.K. Wheaton (2001) 'GM crop data – agronomy and ecology in tandem', *Nature Biotechnology*, 19: 3. [11]

Stewart, G.J. (1992) 'Transformation in the natural environment', in E.M.H. Wellington and J.D. van Elsas (eds.), *Genetic Interactions Among Microorganisms in the Natural Environment*, New York: Pergamon Press, pp. 216–34. [3]

Stoger, E., S. Williams, P. Christou, R.E. Down and J.A. Gatehouse (1999) 'Expression of the insecticidal lectin from snowdrops (*Galanthus nivalis* agglutin; GNA) in transgenic wheat plants: effects on predation by the grain aphid *Sitobion avenae*', *Molecular Breeding*, 5: 65–73. [8]

Szafranski, P., C.M. Mello, T. Sano, C.L. Smith, D.L. Kaplan and C.R. Cantor (1997) 'A new approach for containment of microorganisms: dual control of Streptavidan expression by antisense RNA and the T7 transcription system', *Proceedings of the National Academy of Sciences, USA*, 94: 1059–63. [3]

Tabashnik, B.E. (1994) 'Evolution of resistance to *Bacillus thuringiensis*', *Annual Review of Entomology*, 39: 47–9. [4]

Tabashnik, B.E., Y.-B. Liu, N. Finson, L. Masson and D.G. Heckel (1997) 'One gene in diamondback moth confers resistance to four *Bt* toxins'. *Proceedings of the National Academy of Sciences*, USA, 94: 1640–4. [4]

Talbot, H.W., D.K. Yamamoto, M.W. Smith and R.J. Seidler (1980) 'Antibiotic resistance and its transfer among clinical and non-clinical *Klebsiella* strains in botanical environments', *Applied and Environmental Microbiology*, 39: 97–104. [3, 8]

Tapp, H., L. Calamai and G. Stotzky (1994) 'Adsorption and binding of the insecticidal proteins from *Bacillus thuringiensis* subsp. *kurstaki* and subsp. *tenebrionis* on clay minerals', *Soil Biology and Biochemistry*, 26 (6): 663–79. [7]

Tapp, H. and G. Stotzky (1998) 'Persistence of the insecticidal toxin from *Bacillus thuringiensis* subsp. *kurstaki* in soil', *Soil Biology and Biochemistry*, 30 (4): 471–6. [7]

Tepfer, M. (1993) 'Viral genes and transgenic plants', *Biotechnology*, 11: 1125–32. [4]

Tepfer, M. and M. Jacquemond (1996) 'Sleeping satellites: a risky prospect', *Nature Biotechnology*, 14: 1226. [4]

Teycheney, P.J. and M. Tepfer (1999) 'Gene flow from virus-resistant transgenic crops to wild relatives or to infecting viruses', in P.J.W. Lutman (ed.), *Gene Flow and Agriculture: Relevance for Transgenic Crops*, Farnham: British Crop Protection Council, pp. 191–6. [4]

Thompson, C.E., G.R. Squire, G. Mackay, J.E. Bradshaw, J. Crawford and G. Ramsay (1999) 'Regional patterns of gene flow and its consequences for GM oilseed rape', in P.J.W. Lutman (ed.), *Gene Flow and Agriculture: Relevance for Transgenic Crops*, Farnham: British Crop Protection Council, pp. 95–100. [6]

Thompson, H.V. and C.M. King (1993) *The European Rabbit: The History and Biology of a Successful Coloniser*, Oxford: Blackwell. [5]

Tiedje, J.M., R.K. Colwell, Y.L. Grossman, R.E. Hodson, R.E. Lenski, R.N. Mack and P.J. Regal (1989) 'The planned introduction of genetically engineered organisms: ecological considerations and recommendations', *Ecology*, 70: 298–315. [11]

Timmons, A.M., E.T. O'Brien, Y.M. Charters, S.J. Dubbels and M.J. Wilkinson (1995) 'Assessing the risks of wind pollination from fields of genetically modified *Brassica napus* spp. *oleifera*', *Euphytica*, 85: 417–23. [6]

Timmons, A.M., Y.M. Charters, J.W. Crawford, D. Burn, S.E. Scott, S.J. Dubbels, N.J. Wilson, A. Robertson, E.T. O'Brien, R.G. Squire and M.J. Wilkinson (1996) 'Risks from transgenic crops', *Nature*, 380: 487. [6]

Todd, K. (2001) *Tinkering with Eden: A Natural History of Exotics in America*, New York: W.W. Norton. [5]

Tomiuk, J. and V. Loeschcke (1993) 'Conditions for the establishment and persistence of populations of transgenic organisms', in K. Wöhrmann and J. Tomiuk (eds.), *Transgenic Organisms: Risk Assessment of Deliberate Release*, Basel: Birkhäuser Verlag, pp. 117–33. [5]

Tomlins, C.D.S. (ed.) (1997) *The Pesticides Manual*, 11th Edition, London: British Crop Protection Council. [4]

Tonsor, S.J. (1985) 'Leptokurtic pollen-flow, non-leptokurtic gene-flow in a wind-pollinated herb, *Plantago lanceolata* L.', *Oecologia*, 67: 442–6. [6]

Tudge, C. (1993) *The Engineer in the Garden*, London: Jonathan Cape, pp. 21–2. [11]

Tumlinson, J.H., T.C.J. Turlings and W.J. Lewis (1992) 'The semiochemical complexes that mediate insect parasitoid foraging', *Agricultural Zoology Reviews*, 5: 221–52. [7]

Turlings, T.C.J., J.H. Tumlinson and W.J. Lewis (1990) 'Exploitation of herbivore-induced plant odors by host-seeking parasitic wasps', *Science*, 250: 1251–3. [7]

Tyldesley, J.B. (1973) 'Long-range transmission of tree pollen to Shetland I–III', *New Phytologist*, 72: 175–90, 691–7. [9]

Tynan, J.L., M.K. Williams and A.J. Conner (1990) 'Low frequency of pollen dispersal from a field trial of transgenic potatoes', *Journal of Genetics and Breeding*, 44: 303–6. [6]

Umbeck, P.F., K.A. Barton, E.V. Nordheim, J.C. McCarty, W.L. Parrot and J.N. Jenkins (1991) 'Degree of pollen dispersal by insects from a field test of genetically engineered cotton', *Journal of Economic Entomology*, 84: 1943–50. [6]

USDA (United States Department of Agriculture) (1999) 'The performance of field-released transgenic crops', *USDA Economic Research Service Online*, <http://www.ers.usda.gov/> [4]

Vandermeer, J. (1989) *The Ecology of Intercropping*, Cambridge: Cambridge University Press. [12]

van Veen, J.A., P.J. Kuikman and J.D. van Elsas (1994) 'Modelling microbial interactions in soil: preliminary considerations and approaches', in M.J. Bazin and J.M. Lynch (eds.), *Environmental Gene Release: Models, Experiments and Risk Assessment*, London: Chapman & Hall, pp. 29–46. [3]

Vernadsky, W.I. (1929) *La Biosphere*, Paris: Alcan. [2]

Vidal, J. (1998a) 'Genetically altered crops "could wipe out farmland birds"', *Guardian*, 8 July. [7]

Vidal, J. (1998b) 'Mr Terminator ploughs in', *Guardian (Society)*, 15 April. [8]

Vidal, J. (1999) 'The Lord confronts the brothers Brigham', *Guardian*, 27 July. [11]

Vidal, J. (2001) 'Geneticist proposes "third way" on GM crops', *Guardian*, 19 April. [12]

Vigouroux, Y., H. Darmency, T. Gestat de Garambe and M. Richard-Molard (1999) 'Gene flow between sugar beet and weed beet', in P.J.W. Lutman (ed.), *Gene Flow and Agriculture: Relevance for Transgenic Crops*, Farnham: British Crop Protection Council, pp. 83–8. [6]

Wachbroit, R. (1991) 'Describing risk', in M.E. Levin and H.S. Strauss (eds.), *Risk Assessment in Genetic Engineering*, New York: McGraw Hill, pp. 368–77. [11]

Walker, P. (ed.) (1995) *Larousse Dictionary of Science and Technology*, London: Larousse, p. 472. [1]

Wan, M.T., R.G. Watts and D.J. Moul (1989) 'Effects of different dilution water types on the acute toxicity to juvenile Pacific salmonids and rainbow trout of glyphosate and its formulated products', *Bulletin of Environmental Contamination and Toxicology*, 43: 378–85. [4]

Wang, Z., D.L. Crawford, T. S. Magnuson, B.H. Bleakley and G. Hertel (1991) 'Effects of bacterial lignin peroxidase on organic carbon mineralization in soil, using recombinant *Streptomyces* strains', *Canadian Journal of Microbiology*, 37: 287–94. [3]

Warwick, H. (1999) 'The next GM threat: Frankenstein forests', *The Ecologist Online*, <http://www.ecologist.org/> [9]

Watkinson, A.R, R.P. Freckleton, R.A. Robinson and W.J. Sutherland (2000) 'Predictions of biodiversity response to genetically modified herbicide-tolerant crops', *Science*, 289: 1554–7. [7]

Webb, V. and J. Davies (1994) 'Accidental release of antibiotic-resistance genes', *Trends in Biotechnology*, 12: 74–5. [8]

Welford, R. (1997) *Hijacking Environmentalism: Corporate Responses to Sustainable Development*, London: Earthscan, pp. 16–39. [13]

Wilkinson, J.E., D. Twell and K. Lindsey (1997) 'Activities of CaMV 35S and *nos* promoters in pollen: implications for field release of transgenic plants', *Journal of Experimental Botany*, 48: 265–75. [7, 8]

Wilkinson, M.J., I.J. Davenport, Y.M. Charters, A.E. Jones, J. Allainguillaume, H.T. Butler, D.C. Mason and A.F. Raybould (2000) 'A direct regional scale estimate of transgene movement from genetically modified oilseed rape to its wild progenitors', *Molecular Ecology*, 9: 983–91. [6]

Williamson, M. (1991) 'Biocontrol risks', *Nature*, 353: 394. [3]

Williamson, M. (1996) *Biological Invasions*, London: Chapman & Hall. [3, 5]

Wilson, E.O. (1975) *Sociobiology: The New Synthesis*, Cambridge, Massachusetts: Harvard University Press. [2]

Wilson, E.O. (1992) *The Diversity of Life*, Harmondsworth: Penguin. [2]

Woods, F.W. and K. Brock (1964) 'Interspecific transfer of Ca-45 and P-32 by root systems', *Ecology*, 45: 886–9. [2]

Wraight, C.L., A.R. Zangerl, M.J. Carroll and M.R. Berenbaum (2000) 'Absence of toxicity of *Bacillus thuringiensis* pollen to black swallowtails under field conditions', *Proceedings of the National Academy of Sciences, USA*, 97 (14): 7700–3. [7]

Ye, G.N., H. Daniell and J.C. Sanford (1990) 'Optimization of delivery of foreign DNA into higher plant chloroplasts', *Plant Molecular Biology*, 15: 809–19. [8]

Ye, X., S. Al-Babili, A. Klöti, J. Zhang, P. Lucca, P. Beyer and I. Potrykus (2000) 'Engineering the provitamin A (ß-carotene) biosynthetic pathway into (carotenoid-free) rice endosperm', *Science*, 287: 303–7. [4, 12]

Yenish, J.P., T.A. Fry, B.R. Durgan and D.L. Wyse (1997) 'Establishment of common milkweed (*Asclepias syriaca*) in corn, soybean, and wheat', *Weed Science*, 45: 44–53. [7]

Zhang, P., M. Hayat, C. Joyce, L.I. Gonzalez-Villasenor, C.M. Lin, R.A. Durham, T.T. Chen and D.A. Powers (1990) 'Gene transfer, expression and inheritance of pRSV-rainbow trout GH-cDNA in the common carp *Cyprinus carpio* (L.)', *Molecular Reproduction and Development*, 25: 3–13. [10]

Zhou, X., Y. Liu, L. Calvert, C. Munoz, G.W. Otim-Nape, D.J. Robinson and B.D. Harrison (1997) 'Evidence that DNA-A of a geminivirus associated with severe mosaic disease in Uganda has arisen by interspecific recombination', *Journal of General Virology*, 78: 2101–11. [4]

Zhu, Y., H. Chen, J. Fan, Y. Wang, Y. Li, J. Chen, J.-X. Fan, L. Hu, H. Leung, T.W. Mew, P.S. Teng, Z. Wang and C.C. Mundt (2000) 'Genetic diversity and disease control in rice', *Nature*, 406: 718–24. [12]

Zilinskas, R.A. (1998) 'Analysis of the ecological risks associated with genetically engineered marine micro-organisms', in R.A. Zilinskas and P.J. Balint (eds.), *Genetically Engineered Marine Organisms: Environmental and Economic Risks and Benefits*, Norwell, Massachusetts: Kluwer Academic, pp. 95–138. [3]

Index

A/F Protein, Massachusetts, 139
ActionAid, 171
Advanced Genetic Sciences (AGS), 19
Advanta Seeds, 84-6
Advisory Committee on Novel Foods and
     Processes (ACNFP), UK, 107
Advisory Committee on Releases to the
     Environment (ACRE), UK, 77
Africa, edible plant varieties, 157
Agracetus (Monsanto), 162
Agriculture and Environment
     Biotechnology Committee, UK, 154
albumin, 52
alfalfa, 27
alien species, 54-5
allelochemicals, 89
allergens, 79, 98, 100, 178
American Cyanamid, 35
Antibiotic-resistance, 60, 106-9
aphids, 87-8, 105
apomixis, 116; crops, 159; technology, 117
aquaculture, 132-40; escaped fish, 137
ArboGen, 121, 123
Argentina, 168; transgenic crops, 2
arthropods, transgenic, 145-6
Asda, 85
(*?pp.)Asilomar biosafety conference,
AstraZeneca, 51, 115, 159, 163
atmosphere, chemical composition, 20
Austin, Thomas, 56
Aventis CropScience, 35, 78-79, 81, 84,
     99, 100, 151, 160

Bacillus thuringiensis, 42
baculoviruses, modified, 28
Bailey, M.J., 24
bananas: transgenic, 125; vaccine produc-
     tion, 52
Barker, Fred, 151
Bee Farmers Association, 99
Beet Necrotic Yellow Vein Virus, 60

Berenbaum, May, 93
Bergelson, Joy, 74
Beyer, Peter, 51
biocontrol, 57
biodiversity, 16, 36-7, 128; decline, 39,
     153, 156, 158; indicators, 150; tropical,
     17
biofilms, 29
biogeochemical cycles, 9, 17, 25, 27, 32
bioinsecticides, 28
biome, 10-11
'biopiracy', 161
bioplastics, 48
biopol, 50
'bioprospecting', 162
bioremediation, 22, 31-2
Biosafety Protocol, 180, 182
biosphere, concept, 8, 10
Birch, Nick, 88
birds: population, 94-6; transgenic, 146
black grass, 39
black swallowtail butterfly, 93-4
bolters, 72
brassicas, 63, 66, 89
Brazil, 168
Brazil nuts, 79
British Petroleum (BP), 130
Bromoxynil herbicide, 35, 39-40
Bt: bacteria, 97 crops, 43; pollen, 93;
     toxins, 96, 98; transgenes, 88
Bush, George W., 130

California: Asilomar biosafety conference,
     30-1; ice-minus release, 20; University
     of, 19
Canada, 135; National Farmers' Union, 81
canola, 34, 50, 80-1, 170
Capra, Fritjof, 9
'carbon credits', 129-30
carbon cycle, 183
carbon dioxide, 24, 49; sinks, 129-30

carrots, 57
cassava, 46
catfish, 133
Caulerpa taxifolia, 55
Chile, 129
China, 47, 129; soyabeans, 70; transgenic
    rice, 2, 71
chitinase, 97
chloroplast, 117-19
Ciba-Geigy, 160
CIMMYT, Mexico, 114, 156
CIP, Peru, 114
cleavers, 39
cleistogamy, 110
Climate Change Convention 2000, The
    Hague, 130
clones, transgenic, 147
clouds, 20-21
coffee, 171
Commonwealth Scientific and Industrial
    Research Organization (CSIRO),
    Australia, 56
competitive exclusion principle, 15
conjugation, 29
conservation, *in situ*, 159
Consultative Group on International
    Agricultural Research (CGIAR), 114,
    161
consumers, food chains, 12
contamination, threshold, 86
convective currents, 126
Convention on Biological Diversity, 159
corn earworm, 119
Cornell University, 91-3
cotton, 2, 35, 38-9, 49, 66, 180; Bollgard,
    44-5; bromoxynil-resistant, 40; *Bt*, 43;
    Roundup Ready, 41; transgenic pollen,
    78
cowpea, 42, 46, 98
Crawford, Donald, 24
Crawley, Mick, 59-60
cropping systems, 156
Crops: *Bt*, 44; 'marginal', 158; seed saving,
    114; transgenic, 2, 13, 27, 141-2, 152,
    155, 164, 169, 183; transgenic trial, 99;
    yield increase, 27
cross-pollination, 77, 83, 153; aspens, 127
crown gall disease, 29
Cyanamid, 81

Darwin, Charles, 8
decomposers, food chains, 12
'deep ecology', 9
Delta and Pine Land Company, 112, 115
desertification, 48
Devlin, Robert, 135
dioxins, 140
DNA: naked, 30; universality, 5
drought, 61; resistance, 71, 144
drug patenting, 163
Dundee, University of, 124
DuPont, 35, 160
Dutch elm disease, 124

East Anglia, University of, 20, 95
ecological genomics, 183
ecology, science of, 6, 8
ecosystems, 11; agricultural, 17
English Nature, 85, 94
Environmental Protection Agency (EPA),
    USA, 22-3, 40, 78, 92
enzymes, 3-5, 21, 24, 31, 34-5, 49, 52, 97,
    108-9, 132
'escape', 183
Essex, University of, 167
ethanol, 25, 105; production, 79
European Commission's Standing
    Committee for Foods, 77
European Community Biotech Program,
    51
European corn borer, 44-5, 78, 89, 92,
    109
European Union (EU), 99, 108; precau-
    tionary principle, 173-5, 181; Scientific
    Committee on Plants, 86
eutrophication, waterways, 37
Ewen, Stanley, 104
Exxon Valdez, 21

farm-scale evaluations/trials, UK 141, 147-
    54
farmer suicides, 38
fat hen, 95
fecundity, 11
Federal Environmental Agency, Vienna,
    98
fermenters, 31
fertilization, post- barriers, 110
field trials, 141, 177

fish, 179; freeze-tolerance, 15, 134, 139; toxins to, 37; transgenic, 132-40, 183
Fish and Wildlife Service, USA, 38
Fisher, R.A., 8
Flavr Savr tomato, 104
Fletcher Challenge Forests, 121
flowering plants, 62
food: chains, 8, 12; contamination, 78; labelling, 77; webs, 12
ForBio, 122, 123
forests: area, 121; diversity decline, 128
France, crop evaluations, 154
freeze tolerance, fish, 15, 134, 139
Freiburg, University of, 51
Friends of the Earth (FOE), 76, 94, 98, 152-3; USA, 79
fruit trees, modification, 124
functional foods, 48, 50, 52
Fundación Chile, 122
Fungi: hyphae, 13; toxic effects on, 24

Gaia hypothesis, 9,20
garden pea, 43
genes: banks, 159; Bt, 119; constructs, 113, 132-3; fish, 133; flow, 62-5, 70-1, 83, 102, 109, 116-17, 119, 125, 142, 148, 158-9; 'hunting', 162; silencing, 5, 103, 143; stacking, 144; suicide, 31; terminator, 115
Genesis Research and Development, 121
genetic engineering, definition, 3
Genetic ID, of Iowa, 85
'genetic pollution', 7, 62, 72, 81, 155
GeneWatch, 119
Germany, 155
Ghana, edible plant varieties, 157
global warming, 129
glufosinate herbicides, 34, 39, 59, 67, 83, 100, 151
glyphosate herbicides, 34, 36-7, 40, 80, 83, 124, 151; health incidents, 38-9; resistance, 118
goat grass, 40, 73
Golden Rice (see rice), 51, 165
Gray, A.J., 66
Green Revolution, 1, 49, 156-7, 159, 164, 167, 170
Greenpeace, 50, 151-2
growth hormones, 133, 138; beef, 182; fish, 134-6; mammals, 147
growth promoters, 107

Haeckel, Ernst, 8
Hamilton, Bill, 8, 20-1
Hammond, Edward, 115
Henry Doubleday Research Association, 78
Hepatitis B, vaccine, 52, 104, 125
herbicide-resistance, 6, 33-5, 36, 38, 41, 53, 60-1, 73, 80-1, 83-4, 87, 95-6, 99, 102, 109, 118, 142-4, 149, 154-5, 175
'hidden costs', 164
Hilbeck, Angelika, 89
Hoechst, 35, 160
holistic models, 145
honey, contamination, 98-100
honeybees, 9, 68, 76, 97-101, 141
Huether, Tony, 80
Hutchinson, Evelyn, 14
hybrids: barrier method, 119; crop-weed, 70-5; fish, 137; hybridization, 64-8, 111; 'vigour', 69, 73; weedy, 63; wheat, 156

ice-minus, 19-21
Idaho, University of, 24
IITA, Nigeria, 114
Illinois, University of, 93
imidazolinone herbicides, 35
Imperial College, London, 59
India, 159; edible plant varieties, 157; ICRISAT, 115; illegal planting, 70; patents policy, 161-2; precautionary principle, 181
Indonesia, 129; transgenic trees, 122
Industrial Revolution, 1
Ingham, Elaine, 26
Innsbruck, University of, 21
insects, pollinating, 65, 97
insect-resistance, 144
insecticidal protein, 13
insecticide spraying, reduction, 43
Institute of Virology and Environmental Microbiology, 28
insurance companies, 170
Integrated Pest Management (IPM), 42, 89
intellectual property rights, 81, 108, 112, 121, 161, 163, 171, 179

intercropping, 168
Interlink Associates, USA, 122
International Centre for the Improvement
   of Maize and Wheat (CIMMYT), 156
International Chamber of Commerce,
   Charter for Sustainable Development,
   181
International Coffee Technologies Inc, 116
International Paper Corporation, 121
International Rice Genome Sequencing,
   163
International Rice Research Institute
   (IRRI), 71, 166
introgression, 62
invasion, 54, 178; animal, 56; ecology, 16;
   prediction, 57-8
island biogeography theory, 16
ISSR analysis, 66

Japan, 52, 79, 155
Jena, University of, 100
Jesse, Laura Hansen, 91
Jørgensen, Rikke Bagger, 67
Johnsongrass, 72
juries, UK, 151, 152

Kaatz, Hans-Heinrich, 100
Kudzu vine, 55
Kanamycin, 107
Kenya, 168; precautionary principle, 181
Kermicle, Jerry, 110
keystone species, 17
'killer applications', 51
Klebisella bacteria, 25-6, 35
Kraft taco shells, contamination, 79
Kyoto, Climate Change Convention, 129

Lactobacillus plantarum, 26
laboratories, containment, 31
lacewings, 89
ladybirds, 87
landraces, 158-9, 171
leaching, 71
lectin gene, 42-3, 87-8, 104-5
Leicester, University of, 100
Lepidoptera, British, 28
licencing agreements, 171
life science corporations, 163
Lightfoot, David, 49

lignin, 24-6, 122-4; overproduction, 41
Lindow, Steve, 19
Lenton, Tim, 20-1
Loch Fyne, Scotland, 138
Losey, John, 91
Lovelock, James, 20
lucerne, 57

MacKay, Andrew, 82
Maclean, Norman, 4
maize, 2, 35, 41, 49-50, 58, 60, 64, 66, 70,
   76, 79-80, 93, 110-11, 116, 148-51,
   156, 168, 180; *Bt*, 43, 45, 89-90, 109,
   174; corn leaf blight, 157; fodder, 78;
   hybrid barrier method, 119; organic,
   154; pollen, 76, 91; StarLink, 78-9;
   transgenic seed, 77; USA, 97, 170
malaria, spread, 129
mammals, transgenic, 147
mannose, 109
marine bacteria, 30
marker and promoter genes, 3, 102-6, 108,
   113, 132-3, 143-4, 146, 177
mathematical models, 8, 95
Mayer, Sue, 119
Meacher, Michael, 85
Melchett, Lord, 151-2
Mellon, Margaret, 142
mergers, corporate, 156, 159-60, 163
metabolic systems, limit, 135
methylation, 143
Mexico, illegal planting, 70
'Miami bloc', 180, 182
microorganisms, 6-7, 12, 19-22, 29-30,
   106; self-destruct systems, 31
milkweed, 90-2
mitochondria, 117
Monaco Oceanographic Museum, 55
monarch butterfly, 90-2
monoculture: transgenic crops, 164;
   vulnerability, 156-7
Monsanto (Pharmacia), 22, 34-5, 37, 41,
   43-5, 50, 81, 83, 92, 109, 111-12, 115,
   121-3, 151-2, 162-4, 172; 'crop police',
   82; Technical Use Agreements, 80;
   transgenic seed sales, 160
Monterey pine, 122
Mosaic Viruses, 4, 46, 102
mosquitoes, transgenic, 146

multinational corporations (MNCs), 31,
114, 160
mutualism, 13
mycorrhizal fungi, 13, 23, 25, 38, 56

naked DNA, 30
National Pollen Research Unit, Worcester,
98
National Institute of Agricultural Botany
(NIAB), 76, 83
National Seed List Registration, 148
natural enemies, 87-8
New Zealand, earthworm, 37
niche, concept, 14-15
Nile perch, 54-5
nitrogen-fixation, 48-9
Novartis, 51, 91, 93, 109, 115, 159-60,
163, 174
nuclei, 118
nutraceuticals, 51

oats, 72
Obrycki, John, 91
Odum, Eugene, 20
oil slicks, 21
oilseed rape (canola), 2, 10, 34-5, 39, 46,
50, 57, 61, 66-7, 70-3, 80, 85, 90, 97-
100, 119, 142-3, 149-51, 153-4, 163;
Bt, 60; contamination, 84; growing
bans, 75; male-sterile, 68; planting, 69;
pollen, 65; Varietal Associations, 83;
weedy, 73; wild, 59, 74
oleic acids, 49
oligopoly, corporate, 161-3
Oliver, Melvin, 115
Oregon Primate Research Centre, 108
Oregon State University, 25
organic food, 75-6; crops, 78, 151; farmers,
7, 77, 110, 142, 153, 169
organochlorines, 140
outbreeding, 126
outcrossing, 70
Owusu, Rachel, 129
Oxford, University of, 20
oxygen, generators, 129

parasites, 89; parasitic wasp, 89-90
patents, 109; applications, 161; crop
seeds, 160; Indian policy, 162;

pharmaceutical, 163
pest-resistance, 42-3
pesticides: conventional, 92; poisoning,
164; 'treadmill', 87
Pharmacia and Upjohn (see also
Monsanto), 35, 121, 160
pheromones, 89
phloem, 105
phosphinothricin (PPT) herbicides, 35
photosynthesis, 48-9, 117
pink bollworm, 45; caterpillars, 44;
sterility, 146
Pioneer Hi-Bred, 111, 160; cultivars, 93
Planned Release of Selected And Modified
Organisms, 59
Plant Breeders' Rights (PBRs), 111, 160
plant succession, concept, 8
plantations, 128-9
plasmids, 29-30, 106
plastics, 50
pleiotropic effects, 144
PMI enzyme, 109
pollen, 62, 110, 113, 141, 183; *Bt*, 93;
counts, 68; dispersal, 58, 64, 125;
fertility, 67; gene flow, 63; longevity,
65; maize, 76, 91; pollinators, 13;
transgenic, 9, 78; trees, 127
pollution: 'genetic', 173; fish farms, 136
polychlorinated biphenyls (PCBs), 22-3,
140
poplars, 124
Poppy, Guy, 90
populations, 11, 15, 54; bird, 94-6
Positech, 109
potatoes, 43, 60, 66, 69-70, 88, 104, 156;
blight, Ireland, 157; high-starch, 50;
virus-resistant, 170
Potrykus, Ingo, 51
poverty, 170
precautionary principle, 180-2
'predictive ecology', 170
Pretty, Jules, 167
primate, first transgenic, 108
privatization, genetic resources, 2
producers, food chains, 12
promoter genes (see marker and promoter
genes)
proteinase inhibitors, 42-3, 98
proteins, anti-freeze, 134-5, 138-9; *Bt*, 79

protoplast fusion, 4
Pseudomonas syringae, 19, 21-4, 26, 30, 34
Pusztai, Arpad, 104

rabbits, myxomatosis, 56
RAFI, 115
Rayor, Linda, 93
Reading, University of, 119
recombinant DNA, 2
reductionism, 145
refuge: ecological, 54; size, 44; strategy, 45
reporter genes, 108
reproductive fitness, 58, 60, 73
Research Foundation for Science, Technology and Ecology, 164
Research Seeds of St Joseph, Missouri, 27
resistance management, 44, 124
Rhône-Poulenc (Aventis), 35, 40, 160
rhododendron cultivars, 55
rhodopsin, 183
ribosome inhibitor protein, 113
rice, 65; Basmati, 161; *Bt*, 71; genomes, 163; Golden, 51-2, 144, 165-6; self-pollinating, 70; sticky varieties, 157; varieties, 164
RiceTec, Texas, 161
risk assessment, 17, 31, 141-2, 144-5, 148, 150, 152, 153-4, 178, 183; UK, 149
risks, identification, 18
Risø National Laboratory, Roskilde, 67
Rissler, Jane, 142
Rockefeller Foundation, 51, 166
rotation time, trees, 121-2
Rothamsted Experimental Station, England, 90
Roundup Ready, crop lines, 35, 83
Rowett Institute, Aberdeen, 105
Royal Society, 105
Royal Society for the Protection of Birds (RSPB), 94
rubisco, 49
Rural Advancement Foundation International (RAFI), 112

Safeway, 85
saline tolerance, 48, 61, 71, 144
Salmonella bacteria, 146
salmonids, 133, 138-9
Sandoz, 160

satellite RNA, 47
Sattler, Birgit, 21
Schmeiser, Percy, 81-3
Schuler, Tanja, 90
Schull, George, 111
Schwarz, Mike, 152
Scott, Susan, 119
Scottish Crop Research Institute (SCRI), 67, 74, 88, 99
Seed Crushers' and Oil Producers' Association, 85
seeds: companies, 111-12, 160; contamination, 84-5; crops, 63; dispersal, 63; inadvertent spread, 59; production, 175; Roundup Ready, 82; saving, 114; sterile technology, 113; transgenic crop, 180; viable, 62
self-fertilization, 110
semiochemicals, 89
Sequential Comparison Index, 16
Sharpes International Seeds, 76
Shell, 130
Shetland Islands, 126
Shiva, Vandana, 164
silage, 26
Silvagen, Canada, 122
soapwort plant, 113
sociobiology, 9
Soil Association, 75-8, 151
soil: fertility, 167; samples, 150
South Africa, precautionary principle, 181
Southern Illinois University, Carbondale, 49
soyabeans, 2, 27, 33, 35-6, 38, 41, 66, 80, 180; Brazil, 41; Roundup Ready, 112; USA, 114
StarLink contamination, 79
starvation, 170
sterile insect technique (SIT), 145-6
sterility, 7; fish, 139; male, 110; seeds, 112; soil, 26; trees, 122, 129
Stoneville Pedigree Seed, Memphis, 40
storm systems, 126
strawberries, 69
Streptomyces lividans, 24, 26, 30
stress: resistance, 48; trees, 127
sugar beet, 35, 37, 60, 63, 66, 70, 72, 83-4, 95, 149-50, 153-4; weed, 59
sulphonylurea herbicides, 35

sunflowers, wild, 70
'superweeds', 84
Supply Chain Initiative on Modified
    Agricultural Crops (SCIMAC), 149,
    151, 154
surfactants, 39
sustainable agriculture, 7, 167-9
Swaminathan, M.S., Research Foundation,
    171-2
Swann Chemical Company (Monsanto), 22
sweet potatoes, 45, 172
Swiss Federal Institute of Technology, 51
switches, genetic, 116
Syngenta, 51-2, 109, 111, 115, 123, 159,
    163-6
systems analysis, 9

taxonomists, 5
'technological fixes', 119
teosinte, 110
'Terminator technology', 112, 114, 119,
    159, 171
Tesco, 85
tetracycline, 113
thale cress, 74, 100; genome, 163
'The Archers', 151
tilapia, 140
tobacco: budworm, 119; transgenic, 100
Tolpuddle Martyrs, 152
tomatoes, 70; necrosis, 47
Toyota, 130
'Traitor technology', 116
transduction, 29
transformation, 30
trees, 179; hybrid transgenic, 127;
    longevity, 128; pollination, 126; propa-
    gation techniques, 122; transgenic, 121,
    123-5, 129-31
trehalose, 48
trials, risk, 148
triazine herbicides, 74, 118
TRIPs (trade-related intellectual property
    rights), 161
trophic levels, 12; tri-trophic interactions,
    89-90
trypsin inhibitor, 42
turmeric, patent, 162

Uganda, 46
United Kingdom, (UK), government, 85;
    small-scale trials, 148
United Nations (UN): Biodiversity
    Conferences, 180; Convention on
    Biological Diversity, 162; Food and
    Agriculture Organization, 165-6
United States of America (USA) 181;
    Department of Agriculture (USDA), 41,
    43, 47, 112, 182; Environmental
    Protection Agency (EPA), 22-3, 40, 78,
    92; military, 115; soyabeans, 33;
    Supreme Court, 22; transgenic crops, 2

vaccines, edible, 52
vectors, 3-5
viruses, 103; evolution, 46-7; plant, 5;
    resistance, 45
Vitamin A, 166; deficiency, 51-2
vitamins, deficiency, 165

Wallace, Henry A., 111
Watkinson, Andrew, 95
Watson, Guy, 76-7
weather patterns, 20
weeds: beet, 72; control, 74; hybrids, 63,
    69; invasion, 142; races, 62, 158; 'shift',
    40; uncontrolled growth, 33
Wells, Stewart, 81
Western Canadian Oilseed Association, 34
Westvaco Corporation, 121
wheat, 61, 63, 73, 111, 114; herbicide-
    resistant, 40; hybrids, 156; 'red rust'
    disease, 157
wild turnip, 67-8, 73-4, 119
Wilkinson, Mike, 119
Wisconsin, University of, 110
wood pulp, 121
World Health Organization (WHO), 165-6
World Trade Organization (WTO), 161-2,
    182
World Wildlife Fund, 129
Wytham Wood, 28

zebra mussel, 55
Zeneca, 123
'zero tillage', 168

Zed Books titles on genetic engineering and biotechnology

Scientific advances in biology, the ability to engage in genetic engineering, the inclusion of genetic material under the protective umbrella of intellectual property rights and the commercial interests of the handful of giant corporations who dominate research and development in these fields have created a potent cocktail of change. This nexus of power and technical capacities raise extraordinarily important issues relating to the ethics of manipulating Nature, the consequences for human health, biodiversity and protection of the environment, and social questions including the implications for Third World farmers and food security in the South. Zed Books is developing a strong list on the social, environmental and ethical dimensions of these questions.

M. Avramovic, *An Affordable Development? Biotechnology, Economics and the Implications for the Third World*

Robert Ali Brac de la Perrière and Franck Seuret, *Brave New Seeds: The Threat of GM Crops to Farmers*

Henk Hobbelink, *Biotechnology and the Future of World Agriculture*

Calestous Juma, *The Gene Hunters: Biotechnology and the Scramble for Seeds*

Calestous Juma, *The New Genetic Divide? Biotechnology in the Age of Globalization*

Stephen Nottingham, *Eat Your Genes: How Genetically Modified Food is Entering Our Diet*

Stephen Nottingham, *Genescapes: The Ecology of Genetic Engineering*

Vandana Shiva and Caroline Moser (eds), *Biopolitics: Perspectives on Biodiversity and Biotechnology*

Brian Tokar (ed.), *Redesigning Life? The Worldwide Challenge to Genetic Engineering*

For full details of this list and Zed's other subject and general catalogues, please write to: The Marketing Department, Zed Books, 7 Cynthia Street, London N1 9JF, UK or email Sales@zedbooks.demon.co.uk

Visit our website at: www.zedbooks.demon.co.uk